南京水利科学研究院专著出版基金　资助出版

Shore Nourishment and Morphodynamic Modelling

人工育滩及其动力地貌模拟

孙　波　著

海洋出版社

2015年·北京

图书在版编目（CIP）数据

人工育滩及其动力地貌模拟＝Shore Nourishment and Morphodynamic ModellingSun Bo：英文/孙波著．—北京：海洋出版社，2015.12
ISBN 978-7-5027-9299-2

Ⅰ.①人… Ⅱ.①孙… Ⅲ.①人工制造-海滩-海岸工程-研究-英文 Ⅳ.①P737.11

中国版本图书馆 CIP 数据核字（2015）第 285794 号

责任编辑：朱　瑾
责任印制：赵麟苏

海洋出版社　出版发行

http：//www.oceanpress.com.cn

（100081　北京市海淀区大慧寺路 8 号）

北京华正印刷有限公司印刷　新华书店北京发行所经销
2015 年 12 月第 1 版　2015 年 12 月第 1 次印刷
开本：787mm×1092mm　1/16　印张：10.75
字数：248 千字　定价：48.00 元
发行部：62147016　邮购部：68038093　总编室：62114335

海洋版图书印、装错误可随时退换

Preface

Shore nourishment is the most valid solution for coast erosion. It has been widely used in West Europe and North America as a soft measure, with respect to other hard mitigation methods such as seawall, groin, offshore breakwater etc. The fate of the sand placed in nourishment site is the key issue for engineering design. Nowadays numerical modelling becomes a regular and the most important tool to evaluate coastal project planning. This book concerns numerical modelling on shore nourishment, taking the Egmond shoreface nourishment in the middle of Dutch coast as an example. The hydrodynamic and sediment transport characteristics of nourished shore was analyzed, furthermore the hydrodynamic and morphodynamic performances of the numerical modelling are particularly focused.

All the computations are based on the Delft3D modelling system. Four different types of coastal model, i. e. one-dimensional horizontal profile model, two-dimensional vertical profile model, two-dimensional horizontal area model, and three-dimensional model, have been tested. A two-dimensional horizontal model is firstly setup and used to implement hydrodynamic calibrations against previous study. Further, profile models which are also based on Delft3D are developed to validate the related transport and morphological factors. These factors are then employed in fully 3D area morphodynamic modelling in the site of interest. The model performances are finally evaluated on the basis of the measured data and the previous results.

Coastal hydrodynamics involves water level changes, tidal currents, wave propagations, and wave-current interactions, which makes the coastal water movement complex. Based on these variables, coastal sediment transport and morphodynamic feedbacks become more complicated. Through a specific example of shore nourishment, this book can provide reference postgraduates and coastal engineers to understand numerical modelling on coastal nearshore processes.

This study was implemented at Delft Hydraulics (now named Deltares), the Netherlands, supervised by Prof. Dr. Ir. L. C. van Rijn and Dr. Ir. D. J. R. Walstra. I appreciated them for sharing knowledge and experiences. I gratefully acknowledge my employer Nanjing Hydraulic Research Institute (China) for helping me to study abroad and supporting to publish this book.

Sun Bo

Contents

Chapter 1 Introduction ··· (1)
  1.1　Background ··· (1)
  1.2　Significance of the proposed research ················· (3)
  1.3　Approach and objectives of study ······················· (4)
  1.4　Outline of the book ·· (4)

Chapter 2 Nourishment Behaviour and Modelling ············ (6)
  2.1　Shore nourishment ··· (6)
  2.2　Hydrodynamic and morphodynamic effects of shoreface nourishment ··· (10)
  2.3　Shoreface nourishment modelling ························ (14)
  2.4　Delft3D Modelling System ································ (17)
  2.5　Implementations of Delft3D system in present study ···· (28)

Chapter 3 Egmond Shoreface Nourishment ······················ (31)
  3.1　Egmond beach ··· (31)
  3.2　Egmond Shoreface nourishment ·························· (35)
  3.3　Morphological evolution analysis ························ (38)
  3.4　Conclusions ··· (46)

Chapter 4 Hydrodynamic Modelling ······························ (48)
  4.1　Computational grids and bathymetries ·················· (49)
  4.2　Tidal schematization and calibration ···················· (53)
  4.3　Wave modelling ·· (64)
  4.4　Wave-current interactions ································· (73)
  4.5　Sediment transport modelling ···························· (77)

4.6   Conclusions ·· (90)
Chapter 5   Morphodynamic Validation ···················· (92)
5.1   Morphological scenario ································· (93)
5.2   Calibration on transport factors ······················ (98)
5.3   Morphodynamic simulations of area model ·········· (112)
5.4   Comparison of profile and area modelling ············ (124)
5.5   Conclusions ·· (126)
Chapter 6   Conclusions and Recommendations ············ (129)
6.1   Conclusions ·· (129)
6.2   Recommendations for future study ····················· (133)
Bibliography ··· (135)
Appendix ··· (139)

Chapter 1

Introduction

1.1 Background

Beaches, transition zones between land and sea, provide a measure of protection to the shore from damage by coastal storms. Their effectiveness as natural barriers depends on their size and shape and on the severity of storms. The Dutch coast is naturally eroding when observed over long enough time spans to average out large seasonal variations.

For hundreds of years the Dutch coastlines near Den Helder and near Hoek van Holland are suffering from erosion, due to eroding capacity of tidal and wave-driven currents in combination with the sediment stirring action of the waves and furthermore the sediment-importing capacity of adjacent tidal basins and estuaries. Near Egmond aan Zee (Section 38 km from Den Helder) the retreat of the coastline was about 100 m between 1665 and 1717 (about 2 m/year). Another 120 m of land was eroded between 1717 and 1864; the church tower collapsed on the beach in 1741 (van Rijn, 1995). A large part of the Dutch mainland lying well below mean sea level is now protected from the sea by dunes. The dunes and therefore the beaches are used to be a moving defence system in a dynamic equilibrium. At present the Dutch coast is fixed in its landward movement because of man-made structures (e. g. buildings), which makes it necessary to prevent the current coastline from erosion.

A number of engineering approaches have been used to counteract the effects of erosion by stabilizing or restoring beaches. Traditional protective measures have included "hard" structures such as seawalls, revetments, groins, and detached breakwaters. These structures can reduce flood hazards, armour the coastline, reduce wave attacks, and stabilize the beach. None of these shore protection structures, however, adds sand to the beach system to compensate for natural erosion. An often-used "soft" method to maintain the coastline is sand nourishment at beach or shoreface. Shore nourishment stands in contrast as the only engineered shore protection alternative that directly addresses the problem of a sand budget deficit, because it is a process of adding sand from sources outside the erosion system. Shore nourishment serves as a sacrificial rather than fixed barrier. The nourishment is inherent in the process itself which is most like that of nature and the consequences of the operation for other stretches of coast nearby are probably the least of all the possible protection methods. So the advantages of this method are the relatively small (negative) effects on adjacent coastlines and the relatively low impact on the ecosystem. Furthermore, shore nourishment is usually the cheapest solution than other "hard" ways.

Shore nourishment has received widespread international attentions in the coastal zone management and in the governments of coastal nations. Along the Dutch coast, beach and dune nourishments were carried out regularly during the period 1964—1992. Most of the material was dredged at locations with a depth larger than 20 m or from other locations outside the area of interest (budget area). Totally, 24 nourishments were carried out below the 10 m NAP (Normaal Amsterdams Peil) line in this period; the total nourishment volume is about 15 million cubic meters over the period 1964—1992. The yearly-averaged nourishment volume is about 370 000 m^3/year for the 28 year-period 1964—1991 up to 2 000 000 m^3/year for the three year-period 1990—1992 (van Rijn, 1995).

All shore protection and beach restoration alternatives are controversial with respect to their effects on coastal processes, effectiveness of performance, and socio-economic value. Although the costs of shore nourishment are generally less than the costs of other man-made shore protection works, it is more often than not temporary. If the nourishment is not continuous, the supply has eventually to be repeated. Advancing the state of practice of shore nourishment requires an improved understanding of project location, complex hydrodynamic processes, shoreline evolutions, prediction, design, cost-benefit analysis, monitoring and so on.

1.2 Significance of the proposed research

Several design and planning questions of shore nourishment relate to the fate of the sand placed offshore the surf zone. Can we economically use shoreface nourishment, and what is the certainty that a constructed submerged feature will move onshore or remain in place? And if it will move, then at what speed? So to exactly predict the development of shore nourishment becomes the key issue before carrying out the projects.

Numerical modelling is one of the strongest and most important tools in nowadays coastal engineering. The use of numerical modelling also allows a reduction in study costs. In practice, state-of-art computer models are one- or two-dimensional (depth-average) and have a limited ability to model many of the important three-dimensional flow phenomena found in nature. The use of straightforward two-dimensional horizontal (2DH) morphological models has become more or less commonplace, especially in relatively large-scale applications in tidal inlets, estuaries and coastal areas. Ongoing increases in the computing power available to coastal engineers have meant that morphological simulations of years to decades have become feasible. Three- or quasi-three-dimensional (3D/Q3D) models were also developed to apply on complicated morphological simulations of coastal areas, but their practical applications are limited. In order to improve the performances of the latest modelling systems, it is necessary to perform a large range of validations based on the appropriate modelling concept and high-resolution in-situ observations.

This book concerns numerical modelling on shore nourishment, taking the Egmond shoreface nourishment as an example. All the computations are based on the Delft3D modelling system. Different types of coastal model including 1DH, 2DV, 2DH and 3D are used. Model performances on coastal hydrodynamics and morphodynamics are tested and evaluated.

1.3 Approach and objectives of study

A numerical modelling of morphological changes due to the nourishment will be performed in the study. On the one hand, the validation of the model is carried out; on the other hand, it aims to predict shoreface nourishment behaviour on large spatial-temporal scales. The state-of-the-art Delft3D software package is applied in this study. It will be run in 2DH mode and in 3D mode as area modelling, and in 1DH/2DV mode as profile modelling. The previous related studies and the in-situ measured data are used to compare with the modelled results.

The objectives of this study include the following terms:

(1) Analysis hydrodynamic and morphodynamic characteristics of the Egmond shoreface nourishment;

(2) Design a morphodynamic model (incorporated with wave model) with Delft3D for modelling of the Egmond shoreface nourishment;

(3) Validate, calibrate and evaluate the hydrodynamic and morphodynamic Delft3D model;

(4) Evaluate the performance of the modelling on shoreface nourishment based on the measured and modelled results;

(5) Obtain a frame of reference for further studies.

1.4 Outline of the book

This book is subdivided into six chapters. Except the present chapter, other parts are outlined as follow.

Chapter 2 briefly introduces the natural processes of shore nourishment and nearshore bar migrations. Some information about the development of shore nourishment modelling is then presented. Finally the Delft3D modelling system is introduced in short.

Chapter 3 introduces the Egmond beach and the shoreface nourishment which are the areas of interest. Based on the data analysis, the morphological development of the areas is demonstrated.

Chapter 4 discusses the setup of the Delft3D model at first. The chapter then starts the calibration and validation modelling on three aspects: tide currents, waves and sediment transports. The modelling includes the schematisation of boundary conditions, the settings on physical and numerical parameters, and the comparisons between the results.

Chapter 5 goes into morphological simulations, based on the hydrodynamic modelling. The simulation scenario is specified firstly. To decrease the computation time costs, a profile model is built to calibrate the related transport factors. Afterwards, the calibrated factors are employed to carry out 3D area modelling on the morphological evolutions. The results finally presented and analysed against the measured data and the previous study.

Chapter 6 draws the conclusions towards the modelling performance on the hydrodynamic and morphodynamic features of the shoreface nourishment. It also gives the recommendations for future study.

Chapter 2

Nourishment Behaviour and Modelling

Shore nourishment has emerged as a favoured remedy for coastal erosion in many locations. Large quantities of sand are introduced to the nearshore or offshore zone, having a significant impact on local current and wave field and then accelerating or decelerating the local morphology evolutions. This chapter briefly discusses the basic natural processes of nourishment areas. The most important transport processes and morphological evolution of nourishment enforced by the local current and local wave field will be explained.

Computer modelling of sediment transport patterns is generally recognized as a valuable tool for understanding and predicting morphological developments. The Delft3D modelling system is a well-known software package for water engineering. The system is applied in this study to simulate the morphological evolutions of offshore nourishment. This chapter introduces the system briefly; some details will be presented in the following chapters incorporating with model setup.

2.1 Shore nourishment

The traditional engineering response to coastal erosion has been to mitigate and where possible prevent erosion by coastal structures such as seawalls, groins and breakwaters. This thought of counteracting instead of working in concert with natural processes is now referred to as "hard engineering". Nevertheless, the long-term monitoring of coastal changes around such structures frequently shows adverse environmental effects, in the vicinity (near field) as well as further away on adjacent shores (far field). Furthermore, the recognition of the necessity for sustainable development of the coastal environment has led coastal engineers to the present interest in developing a "soft engineering" approach (Hamm *et al.*, 2002). Once the beach erosion phenomenon has been defined, by identifying both the causes (through a study on morphological processes) and the local interests (evaluating the different aspects related to safety, recreation, environment, and economy), the careful selection of the protective measures to be adopted is of primary interest. Nowadays shore nourishment, if necessary integrated with some "supporting structures", is an increasingly important option.

A central technique used in the soft engineering approach is shore nourishment. The use of the term "shore nourishment" rather than "beach nourishment" is preferred, since the nourishment location may vary considerably in the cross-shore direction i. e., on the first dune row, at the dune face, on the beach, in the surf zone and/or at the shoreface. Today, periodic nourishment is regarded as an environmentally acceptable method of shore protection and restoration for short-term emergencies (i. e. storm-induced erosion) as well as long-term issues (i. e. structural erosion and sea level rise). The philosophy behind nourishment is based on the consideration that when a stretch of coast is sediment-starved, it could be more appropriate to import sediment and let nature do its job, rather than desperately try to counteract natural forcing factors to keep the remaining sediment.

Considering traditional strategy of using hard engineering, the history of beach nourishment in Europe is pretty recent. Hamm *et al.* (2002) summarises a few basic nourishment facts of Europe and USA before 1990. An inventory of the practices and objectives of beach fill projects from in Europe has listed by Hanson *et al.* (2002). Shore nourishment started in 1950 at Estoril, near Lisbon, Portugal, with a nourishment scheme involving some 15 000 m^3, which was soon followed by another in 1951 on the island of Norderney, Germany. Also, the United Kingdom started nourishment in the early 1950s. France followed in the early 1960s, Belgium and Italy in the late 1960s, the Netherlands in 1970, Denmark in 1974 and Spain in the early 1980s. In the early 21 century, the total annual rate of nourishment in Europe adds up to about 28M m^3, which is about the same volume as that for federal projects in the USA.

Along the Mediterranean coast of Spain, artificial nourishment has become the principal remedial answer to coastal erosion problems. In the Netherlands, Denmark (west coast of Jutland), the United Kingdom and Germany (North Sea coast), coastal protection against flooding by storm surges is a vital issue and has given rise to the development of long-term intervention strategies, which are implemented through follow-up programmes.

Nourishment types can be distinguished by focusing on the cross-shore and the longshore dimensions. The cross-shore dimension is the main factor determining the approach of nourishment intervention; whereas, the longshore dimension is the main factor determining the environment where the nourishment intervention is undertaken (Capobianco *et al.*, 2002). Table 2.1 summarises the types of nourishment approaches, based on the cross-shore location of placement. Table 2.2 summarises the nourishment environment, based on the longshore dimension.

Table 2.1 Types of nourishment approach focusing on cross-shore dimension

Nourishment approach	Description
Dune nourishment	placing all the sand as a dune behind the active beach
Nourishment of subaerial beach	using the nourished sand to build a wider and higher berm above the mean water level
Profile nourishment	distributing the added sand over the entire beach profile
Bar or shoreface nourishment	placing the sand offshore to form an artificial bar

Chapter 2   Nourishment Behaviour and Modelling

Table 2.2   Nourishment situations focusing on longshore dimension

Situation/Environment	Description
Open coast	placing the sand on a relatively open coast
Pocket beach	placing the sand on a completely closed beach
Groin or other structures	placing the sand down-drift of a groin or in the presence of other structures
Tidal inlet	placing the sand on a barrier beach or a spit close to a tidal inlet

In 2011, the Sand Motor project (also called Sand Engine), a unique shore nourishment project in the world, was carried out in the Netherlands. From March 2011 to November 2011, a hook-shaped peninsula in South Holland coast was created by man.

Fig. 2.1 shows the location of the Sand Motor. It filled 21.5×10^6 m^3 of sand, which extended 1 km into the sea and is 2 km wide along the shore. Trailing suction hopper dredgers picked up the sand ten kilometres off the coast and took it to the right place. Two offshore replenishment locations alongside the peninsula are also part of the Sand Motor. The project is a great example of "building with nature" and an experiment in the management of dynamic coastline. It is expected that the filled sand is then moved over the years by the action of waves, wind and currents along the coast. To protect the West of the Netherlands against the sea, the beaches along the coast are artificially replenished every five years, and it is expected that the sand engine will make replenishment along the Delfland Coast unnecessary for the next 20 years. This method is expected to be more cost effective and also helps nature by reducing the repeated disruption caused by dredging and replenishment.

Wind and currents started to change the Sand Motor as soon as it was created. The alterations in its shape have largely matched expectations so far. According to periodical measurements, the Sand Motor has been eroded on the western side, with the sand being deposited to the north and south. As a whole, it has become narrower and longer. The northern side is most changeable: the shapes of the lagoon, channels and sandbanks are constantly altering.

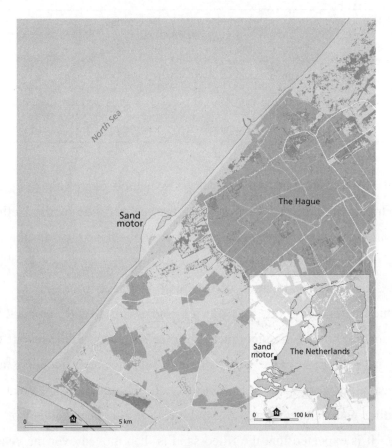

Fig. 2.1 Location of the Sand Motor project.

2.2 Hydrodynamic and morphodynamic effects of shoreface nourishment

A shoreface nourishment (underwater nourishment) can be seen as a submerged structure such as a soft reef berm or a submerged, hard-rock breakwater. Submerged hard-rock structures have been built at many sites around the world. A basic effect is the reduction of wave height in the nearshore zone in the lee of the structure, leading to a more stable shoreline (Browder et al., 2000). If the crest level of a breakwater is relatively low, it does not provide a significant reduction of wave height and has a rather limited impact on the morphological system. A relatively high crest level may result in substantial deposition of sand in the lee zone. Heavy erosion in the gaps between two breakwaters and scouring effects at the shore-side distal ends of the barriers have been observed near rubble-mound breakwaters. There placement of emerged structures by submerged breakwaters has resulted in the disappearance of tombolos and shoreline recession. Thus, additional maintenance nourishments of the beaches are often required. At present, beach nourishment protected by submerged structures is considered the most effective solution for the Italian coasts (Lamberti and Mancinelli, 1996).

The generation of setup currents due to the increased water level in the lee of a submerged breakwater, which is a result of water transport over the breakwater generated by wave breaking, is one of the most important hydrodynamic effects. This surplus water trapped inshore of the submerged breakwater drives currents that flow along paths of least resistance toward the distal ends of the breakwater. The main portion of the water is directed alongshore, producing a longshore current, see Fig. 2.2 (van Duin *et al.*, 2004).

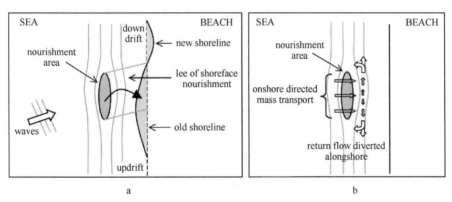

Fig. 2.2 Lee effects of the placement of a shoreface nourishment.

Morphological study of a stable reef berm in relatively deep water along the east coast of Florida shows that the berm does not migrate and is smoothened only slightly (Work and Dean, 1995; Work and Otay, 1996). The wave breaking was infrequent for a water depth larger than about 5 m above the crest and the energy reaching the shoreline was hardly reduced. The shoreline erosion at this beach in Florida (USA) in the lee of the berm was significantly smaller than the erosion occurring outside the lee zone of the berm. This favourable effect was caused by the sheltering effect of the berm. To cause the larger waves to break over the barrier, the crest immersion had to be about half the original water depth (Zwamborn et al., 1970). Although nonbreaking waves also caused a reduction in wave height in the lee of the berm, the wave height reduction for conditions with breaking waves showed much larger reductions. The berm was most effective if the crest width was about eight times the depth of water above the crest. The response of the berm to storm events showed minor erosion at the crest. The beaches protected by the berm benefited greatly.

Observations of active feeder berms showed that these berms mostly move onshore. Some berms remained stable, but no berms moved seaward (Ahrens and Hands, 1998). The morphological behaviour of a feeder berm along the central section of the North Sea barrier island of Terschelling, The Netherlands, was studied by Hoekstra et al. (1996), Spanhoff et al. (1997), and Grunnet (2002). The shoreface nourishment migrated in the dominant alongshore drift direction as well as in the shoreward direction. The nourishment area showed overall erosion, while the landward zone in the lee of the nourishment area showed accretion. The accretion was much larger than the onshore migration, thus sediment must have entered the study area by longshore transport processes due to shielding effects caused by the nourishment area.

Shoreface nourishment can be considered as an artificial offshore bar or a "soft" submerged breakwater. It has similar functions of offshore submerged breakwater, and follows the rule of longshore bar migration as well. Offshore nourishment functions by locally reducing the amount of nearshore wave energy thereby creating a "shadow zone" where longshore transported sediments accumulate. The nourishment reflects or dissipates the incident wave energy and alters the wave direction and height by refraction and diffraction, thereby modifying the local longshore transport, which is so-called "lee effect". Since the nourishment is a soft structure other than hard structures, its own shape will be changed when it counteracts the local hydrodynamic conditions.

Surf zone sandbars protect beaches from wave attacks and are a primary expression of cross-shore sediment transport. The typical beach-bar behaviour on the time scale of the seasons is the offshore-onshore migration cycle, in which the offshore migration of the bar system often happens during the winter season (high waves) and the onshore migration, and beach recovery does during the summer season (low waves). Such seasonal variation resulting in so-called winter and summer profiles are a general characteristic of nearshore morphological behaviour, but the degree of seasonality varies widely. During storms, intense wave breaking on the bar crest drives strong offshore-directed currents (undertow) that carry sediment seaward, resulting in offshore sandbar migration. If the beach morphology is in equilibrium, the offshore migration is balanced by slower onshore transport between storms.

Hoefel and Elgar (2003) schematised the feedbacks that drive sandbar migration:

(a) **Large waves** in storms break on the sandbar, driving a strong offshore-directed current (undertow) that is maximum just onshore of the bar crest. The cross-shore changes (gradients) in the strength of the undertow result in erosion onshore, and deposition offshore of the sandbar crest, and thus offshore bar migration. The location of wave breaking and the maximum of the undertow move offshore with the sandbar, resulting in feedback between waves, currents, and morphological change that drives the bar offshore until conditions change.

(b) **Small waves** do not break on the bar, but develop pitched-forward shapes. Water is rapidly accelerated toward the shore under the steep front face of the waves and decelerates slowly under the gently sloping rear faces. Thus, the time series of acceleration is skewed, with larger onshore than offshore values. The cross-shore gradients in acceleration skewness (maximum on the bar crest) result in erosion offshore, and deposition onshore of the bar crest, and thus onshore bar migration. The location of the peak in acceleration skewness moves onshore with the sandbar, resulting in feedback between waves, currents, and morphological change that drives the bar onshore until conditions change.

2.3 Shoreface nourishment modelling

The adoption of simulation models represents technique to answer theoretical and/or hypothetical questions. They are used most appropriately when a problem under analysis is too complex to be solved by analytical models. Simulation (through models) is quantitative procedure that describes morphodynamic processes by constructing a model and then observing how the model behaves over a series of iterations in order to learn how the process itself might behave (Capobianco et al., 2002). The quantity and quality of data input is critical for establishing the goodness of modelling and of model application. Models should also be used to evaluate existing uncertainties and their effects, particularly while dealing with aspects that are not fully monitored or are not fully known. Models also provide the basis for risk assessment to be undertaken. Thus, it is valuable to run models more than once under a variety of conditions and to undertake sensitivity analyses with respect to key parameters.

Sediment motion in the nearshore is an extremely complex phenomenon. The hydrodynamics of wave and current motions at different time scales including tides and surf-beat frequencies, winds, density-driven currents, turbulence, and the interactions of these processes with the bottom contribute to the near-bottom velocity field responsible for sediment motion. The response of the sediments to the hydrodynamics is even more challenging and less understood. Theoretical treatments are limited by numerous variations in sediment transport processes such as bed and suspended load, ripple and dune migration, sheet flow, and so forth.

Shore nourishment evolution is clearly three-dimensional processes. Both longshore and cross-shore sediment transport are important in different parts of the domain. This is a typically complex coastal case that combines: ① wave-driven longshore and cross-shore currents; ② flow acceleration, deceleration, and curvature; ③ non-equilibrium sediment concentrations due to waves and currents; ④ flooding and drying of the computational cells; ⑤ significant morphological changes.

Nonetheless, those tasked with solving problems and making decisions in this environment make the best use of available technology while pursuing a better understanding of the physics involved in order to develop improved methodologies. Analytical and numerical techniques for prediction of shore nourishment behaviour exist at varying levels of sophistication. Computer modelling of sediment transport patterns is generally recognized as a valuable tool for understanding and predicting morphological developments.

Capobianco *et al.* (2002) reviewed the main classes of models currently available. On the basis of the dimensions described by the models, morphological models can be classified to four main types, listed in Table 2.3. Rolevink and Reniers (2012) classified coastal morphology models to three types: ① coastal profile models, where the focus is on cross-shore processes and the longshore variability is neglected; ② coastline models, where the cross-shore profile are assumed to retain their shape even when the coast advances or retreats; ③ coastal area models, where variations in both horizontal dimensions are resolved. These coastal area models are further subdivided into two-dimensional horizontal (2DH) models, which use depth-averaged equations, and three-dimensional (3D) models, which resolve the vertical variations in flow and transport.

Table 2.3 Type of models for shore nourishment

Type of model	Description
Profile Evolution	The model describes the evolution of a cross-shore profile, with possible consideration, in some simplified manner, of the connection with adjacent profiles. Experience with profile evolution models covers mainly the simulation of profile response to extreme events.
Shoreline Evolution	The model describes the evolution of a coast-line, with possible consideration, in some simplified manner, of the connection with other contour lines. Experience with shoreline evolution models covers long-term morphodynamics and the presence of structures.
Multi-layer	Multi-layer models as hybrid combinations of cross-shore and coastline models. Experience with multi-layer models covers mainly scenario evaluation in data-scarce situations.
Quasi 3D	The model describes the whole tri-dimensional evolution of a coastal stretch. Quasi 3D models are largely experimental; they are suitable for the analysis of "initial response" of the coastal system and will most likely be further developed to handle situation with complex morphology.

Numerical modelling is distinguished as short-term event-based and long-term, on the basis of their temporal scale of interest. Short-term event-based modelling focus on the evaluation of the short-term evolution of a disturbance (in the morphology or in the forcing), otherwise long-term modelling focuses on the evaluation of the long-term evolution of a disturbance (in the morphology or in the forcing). On the basis of model's underlying philosophy, they have different types listed in Table 2.4. Roughly speaking, process-based models tend to be appropriately used for short-term scales, while descriptive, equilibrium and behaviour-oriented models are more appropriately used for long-term scales. The essence of behaviour-oriented models, in particular, is the identification of simple parametric mathematical models which exhibit a similar dynamic behaviour as the actual coastal morphology (De Vriend *et al.*, 1993). The model equations are selected because their solution exhibits the suitable behaviour in a certain class of applications. In the application to nourishment planning and evaluation the choice of model structure is led by the prior knowledge about the behaviour of the morphological system subject to the nourishment intervention.

Table 2.4 Type of models for underlying philosophy

Type of model	Description
Process-based	based on the detailed description of the different processes which originate the morphology
Descriptive	based on a classification of observed beach states
Equilibrium	based on the a priori identification of an equilibrium state without describing the way such equilibrium is achieved
Empirical (behaviour-oriented)	based on an empirical (or a behaviour oriented) description of the tendency towards the equilibrium

2.4 Delft3D Modelling System

Coastal Profile and Coastal Area models are the two main generic types of process-based models. Coastal Profile models reflect the physical processes in a cross-shore direction, assuming longshore uniformity. All relevant transport components in the cross-shore direction such as wave asymmetry and the presence of mean cross-shore currents are included. Bed level changes follow from numerical solution of the mass conservation balance. Longshore wave-driven and tide-driven currents and the resulting sediment transport are included in most models. Coastal area models are two- or three-dimensional horizontal models consisting of, and linking, the same set of sub-models of the wave field, the tide-driven and wave-driven flow field, the sediment transport fluxes and the bed evolution (van Rijn et al., 2003). The Delft3D modelling system is an example of such models.

Delft3D is an integrated modelling suite, which simulates two-dimensional (in either the horizontal or a vertical plane) and three-dimensional flow, sediment transport and morphology, waves, water quality and ecology and is capable of handling the interactions between these processes. It has been designed to simulate the hydrodynamic and morphodynamic behaviour of rivers, estuaries, and coasts on time scale of days to years. The software is continuously improved and developed with innovating advanced modelling techniques as consequence of the research work of Delft Hydraulics (Deltares) and to stay world leading. In 2011, the Delft3D flow, morphology and waves modules are available in open source. The software is applied in this book.

The following section gives a general introduction on the Delft3D modelling system, for detailed information see the software technical manual and user manual. In this study, a surf zone wave (SZW) model was also used to perform wave calculations, except for Delft3D-WAVE module. The SZW-model has been integrated into the Delft3D-FLOW module. In the independent Delft3D-WAVE module, two wave models are included, which are HISWA (HIndcast Shallow Water WAves) and SWAN (Simulating WAves Nearshore) respectively. At last, a brief introduction to morphological simulation is also presented as it results from recent advances in ongoing research.

2.4.1 Flow module

The FLOW module is the heart of Delft3D and is a multi-dimensional (2D or 3D) hydrodynamic (and transport) simulation programme which calculates non-steady flow and transport phenomena resulting from tidal and meteorological forcing on a curvilinear, boundary fitted grid or spherical coordinates. In 3D simulations, the vertical grid is defined following the so-called sigma coordinate approach or Z-layer approach. The vertical grid of the sigma coordinate approach consists of layers bounded by two σ-plane (-1, 0), see Fig. 2.3. This means that over the entire computational area, irrespective of the local water depth, the number of layers is constant. As a result a smooth representation of the topography is obtained.

The governing equations of the program consist of the horizontal

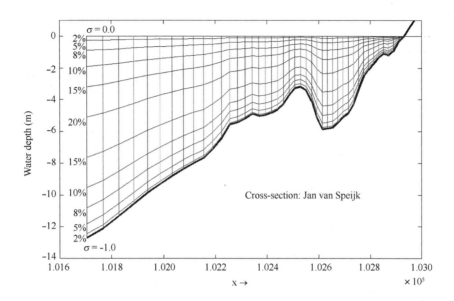

Fig. 2.3 Vertical profile of 3D grid. This figure shows σ-coordinate approach in vertical direction, which means over the entire computational area, irrespective of the local water depth, the number of layers is constant.

momentum equations, the continuity equation, the transport equation, and a turbulence closure model. The shallow water equations can be derived from the more general Navier-Stokes equations, i. e., the momentum balance:

$$\frac{\partial u}{\partial t} + u\frac{\partial u}{\partial x} + v\frac{\partial u}{\partial y} + w\frac{\partial u}{\partial z} - f_{cor}v = \frac{\partial}{\partial x}\left(v_h\frac{\partial u}{\partial x}\right) + \frac{\partial}{\partial y}\left(v_h\frac{\partial u}{\partial y}\right) + \frac{\partial}{\partial z}\left(v_v\frac{\partial u}{\partial z}\right) -$$

$$\frac{1}{\rho}\frac{\partial p}{\partial x} + \frac{W_x}{\rho} \qquad (2.1)$$

$$\frac{\partial v}{\partial t} + u\frac{\partial v}{\partial x} + v\frac{\partial v}{\partial y} + w\frac{\partial v}{\partial z} + f_{cor}u = \frac{\partial}{\partial x}\left(v_h\frac{\partial v}{\partial x}\right) + \frac{\partial}{\partial y}\left(v_h\frac{\partial v}{\partial y}\right) + \frac{\partial}{\partial z}\left(v_v\frac{\partial v}{\partial z}\right) -$$

$$\frac{1}{\rho}\frac{\partial p}{\partial y} + \frac{W_y}{\rho} \qquad (2.2)$$

$$\frac{\partial Uh}{\partial x} + \frac{\partial Vh}{\partial y} + \frac{\partial \eta}{\partial t} = 0 \qquad (2.3)$$

and the mass balance:

$$\frac{\partial \rho}{\partial t} + \frac{\partial \rho u}{\partial x} + \frac{\partial \rho v}{\partial y} + \frac{\partial \rho w}{\partial z} = 0 \qquad (2.4)$$

These equations are valid for any flow type, where the velocity in x, y, and z direction is given by u, v, and w respectively, ρ is the water density, η is the water free surface, the Coriolis parameter (inertial frequency) $f_{cor} = 2\Omega \sin \varphi$, with the earth rotation angular frequency at 2π/day and φ the latitude. In most coastal area horizontal scales are very much larger than vertical scales, so different horizontal and vertical turbulence viscosities v_h, v_v are used. The effect of averaging over the wave motions leads to extra terms in the momentum balance but also in a non-zero time-averaged wave-related pressure (Longuet-Higgins and Stewart, 1962) which is included in the wave forces W_x and W_y. Taking the wave actions into account, U and V represent the depth- and wave-averaged flow velocities.

We assume that the flow is incompressible and density is uniform, so that the mass balance reduce to a volume balance:

$$\frac{\partial u}{\partial x} + \frac{\partial v}{\partial y} + \frac{\partial w}{\partial z} = 0 \qquad (2.5)$$

The vertical pressure gradient follows the so-called hydrostatic balance, i.e., the vertical momentum balance:

$$\frac{\partial p}{\partial z} = -\rho g \qquad (2.6)$$

This can be integrated easily to obtain the pressure at elevation z, for a water level η and an atmosphere pressure p_a:

$$p = p_a + \int_z^\eta \rho g dz \qquad (2.7)$$

A large number of processes included in Delft3D-FLOW (e.g. wind shear, wave forces, tidal forces, density driven flows and stratification due to salinity and/or temperature gradients, atmospheric pressure changes, drying and flooding of inter-tidal flats, etc.) mean that Delft3D-FLOW can be applied to a wide range of river, estuarine, and coastal situations. A number of modifications have recently been incorporated in the FLOW module to account for the three-dimensional effects of waves on the computed flow velocities and turbulent mixing values. In earlier versions of the flow module, the only wave effects included were a breaking wave-induced shear stress at the surface and an increased bed shear stress; in the applied version, recent improvements to compute the wave-averaged currents include major wave-current processes such as wave-induced mass flux, wave-induced turbulence, the effects of streaming and forcing due to wave breaking (Walstra et al., 2001). For a 3D simulation, a turbulence closure model can be used to determine the vertical eddy viscosity and the vertical eddy diffusivity. In the present study, a k-epsilon turbulence closure model was applied.

For sediment transport, a latest approach so-called "online sediment transport" has been implemented in FLOW module. One of the advantages for the use of online sediment transport is the continuous updating of the bed-level and feedback to the hydrodynamics. With the feedback of bottom changes to the hydrodynamic computations, it's possible to execute a full 3D morphodynamic computation with online sediment transport. The online sediment approach allows calculation of morphological changes due to the transport, erosion, and deposition of both cohesive (mud) and non-cohesive (sand) sediments in conjunction with any combination of the above processes. This makes the online sediment version of Delft3D-FLOW especially useful for investigating sedimentation and erosion problems in complex hydrodynamic situations.

The main advantages of this online approach are summarised as: ① three-dimensional hydrodynamic processes and the adaptation of non-equilibrium sediment concentration profiles are automatically accounted for in the suspended sediment calculations; ② the density effects of sediment in suspension (which may cause density currents and/or turbulence damping) are automatically included in the hydrodynamic calculations; ③ changes in bathymetry can be immediately fed back to the hydrodynamic calculations; ④ sediment transport and morphological simulations are simple to perform and do not require a large data file to communicate results between the hydrodynamic, sediment transport, and bottom updating modules.

In the FLOW module the sediment fractions and an optional morphodynamic computation are activated. The influence of waves can be included by running Delft3D-WAVE in coupling with Delft3D-FLOW. More information is presented in Section 2.4.4.

2.4.2 Wave module

In the Delft3D-WAVE module, two wave models are available. These are the second generation stationary HISWA wave model (Holthuijsen *et al.*, 1989) and the third generation spectral SWAN model (Ris *et al.*, 1999; Booij *et al.*, 1999). The SWAN model is the successor of the HISWA model. The main differences between two models with respect to the physics and numeric are:

(1) The physics in SWAN are explicitly represented with state-of-the-art formulations (whereas HISWA uses highly parameterized formulations for the physical formulations);

(2) The SWAN model is fully spectral in frequencies and directions (0° ~ 360°) (whereas the HISWA model is parameterized in frequency, which does not allow for the simulation of multi-modal wave fields);

(3) The wave computations in SWAN are unconditionally stable due to the fully implicit schemes that have been implemented (so, wave propagation in SWAN is not limited to a directional sector of 180° as in the HISWA model);

(4) The computational grid in SWAN has not to be oriented in the mean wave direction (as in the HISWA model).

Several other differences between the two wave models, which may be of importance in practical applications of the Delft3D-WAVE module, are:

(1) SWAN can perform computations on a curvilinear grid (better coupling with the Delft3D-FLOW module);

(2) The wave forces can be computed on the dissipation rate or the gradient of the radiation stress tensor (rather than on the dissipation rate only as in the HISWA model);

(3) Output can be generated in terms of one- and two-dimensional wave spectra in SWAN.

The disadvantage of SWAN is the longer computation time than that of the HISWA model (about 20 times larger) . Due to the increasing of computer power, this disadvantage could be omitted. The wave model applied in this book is SWAN.

In SWAN the waves are described with the two-dimensional wave action density spectrum, even when non-linear phenomena dominate (e. g. , in the surf zone) . The rational for using the spectrum in such highly non-linear conditions is that, even in such conditions it seems possible to predict with reasonable accuracy this spectral distribution of the second order moment of the waves (although it may not be sufficient to fully describe the waves statistically) . The spectrum that is considered in SWAN is the action density spectrum $N(\sigma,\theta)$ rather than the energy density spectrum $E(\sigma,\theta)$ since in the presence of currents, action density is conserved whereas energy density is not. The independent variables are the relative frequency σ (as observed in a frame of reference moving with the current velocity) and the wave direction θ (the direction normal to the wave crest of each spectral component). The action density is equal to the energy density divided by the relative frequency:

$$N(\sigma,\theta) = E(\sigma,\theta)/\sigma \qquad (2.8)$$

In SWAN this spectrum may vary in time and space. The evolution of the wave spectrum is described with the spectral action balance equation which for Cartesian co-ordinates is:

$$\frac{\partial}{\partial t}N + \frac{\partial}{\partial x}C_x N + \frac{\partial}{\partial y}C_y N + \frac{\partial}{\partial \sigma}C_\sigma N + \frac{\partial}{\partial \theta}C_\theta N = \frac{S}{\sigma} \qquad (2.9)$$

The first term in the left-hand side of this equation represents the local rate of change of action density in time, the second and third term represent propagation of action in geographical space (with propagation velocities C_x and C_y in x and y space, respectively) . The fourth term represents shifting of the relative frequency due to variations in depths and currents (with propagation velocity in σ space) . The fifth term represents depth induced and current-induced refraction (with propagation velocity in θ space) . The expressions for these propagation speeds are taken from linear wave theory. The term $S(\sigma,\theta)$ at the right-hand side of the action balance equation is the source term, in terms of energy density representing the effects of wave generation, dissipation and non-linear wave-wave interactions.

The SWAN model is driven by wind and wave boundary conditions and is based on a discrete spectral balance of action density that accounts for refractive propagation of random, short-crested waves over arbitrary bathymetry and current fields. In SWAN, the processes of wind generation, white capping, nonlinear triad and quadruplet wave-wave interaction, bottom dissipation and depth-induced wave breaking are represented explicitly. The numerical scheme for wave propagation is implicit and therefore unconditionally stable at all water depths. To model the energy dissipation in random waves due to depth-induced breaking, a spectral version of the bore-based model of Battjes and Janssen (1978) is used to model bottom-induced dissipation; the JONSWAP (Joint North Sea Wave Project) formulation is applied to compute bottom friction. The formulation for wave-induced bottom stress is modelled according to Fredsøe (1984). Field verifications of the SWAN model have proven its ability in accurately reproducing wave height and period distribution, even in complex coastal areas such as barrier islands and tidal flats (Holthuijsen, 2010).

2.4.3 Surf zone wave model

To decrease the computation efforts, the wave simulation can be carried out by Surf Zone Wave (SZW) model in which wave direction is calculated following Snell's law and wave energy dissipation is calculated by roller model. This implementation has been integrated into the Delft3D-FLOW module, so it is not necessary to use Delft3D-WAVE module in some circumstances. The brief introduction about the roller model is given below, more details see the Delft3D-FLOW User manual.

The roller model utilised the energy balance of the shoaling and breaking random short waves to predict the cross-shore variation of the short-wave energy (and radiation stresses) on the scale of the wave groups, and applied the results as a forcing in the numerical calculation of incident waves. When waves travel across shallow water, the short wave energy balance reads:

$$\frac{\partial E}{\partial t} + \frac{\partial}{\partial x}(EC_g \cos \alpha) + \frac{\partial}{\partial y}(EC_g \sin \alpha) = -D_W \quad (2.10)$$

in which, E is short wave energy, C_g is wave group velocity, α is wave direction, D_W is dissipation of wave energy.

Through the process of wave breaking, the wave energy is reduced and transformed into roller energy E_r. This energy is located in the down-wave region after wave breaking. Spatial variation in the roller also generates forces on the water. The roller energy is rapidly dissipated in shallow regions. The energy that is lost from the organised wave motion is converted to roller energy through the roller energy balance.

$$\frac{\partial E_r}{\partial t} + \frac{\partial}{\partial x}(2E_r C \cos\alpha) + \frac{\partial}{\partial y}(E_r C \sin\alpha) = D_W - D_r \quad (2.11)$$

where C is the wave celerity. The roller energy dissipation D_r is a function of the roller energy E_r:

$$D_r = 2\beta g \frac{E_r}{C} \quad (2.12)$$

Here β is a user-specified coefficient of approximate 0.1 and g the acceleration of gravity. The model needs wave direction and period information from wave input file (named as wavecon.rID). The main purpose of the roller model is to include the roller equations, which leads to a shoreward shift of the wave set-up and the longshore and cross-shore flow.

In the profile modelling of this study, we applied the roller model in stationary mode (combined by Snell's law) with a combination of water level boundaries for offshore boundary and Neumann boundaries for the lateral boundaries. On the other hand, roller model can also be used associated with Delft3D-WAVE. In this situation, the wave direction is obtained from SWAN model, but wave energy dissipation is then simulated by roller model. See (Reniers et al., 2003; Roelvink and Reniers, 2012) for more information about roller model.

2.4.4 Morphodynamic modelling

Sediment transport (both suspended and bed total load) and morphological changes for an arbitrary number of cohesive and non-cohesive fractions are integrated in the Delft3D-FLOW module. Both currents and waves act as driving forces and a wide variety of transport formulae have been incorporated. For the suspended load the morphodynamic process connects to the 2D or 3D advection-diffusion solver of the FLOW module; density effects may be taken into account. An essential feature of the morphodynamic process is the dynamic feedback with the FLOW and WAVE modules, which allow the flows and waves to adjust themselves to the local bathymetry and allows for simulations on any time scale from days (storm impact) to centuries (system dynamics). It can keep track of the bed composition to build up a stratigraphic record. The morphodynamic process may be extended to include extensive features to simulate dredging and dumping scenarios.

Conventional morphological modelling is carried out using a morphodynamic feedback loop and consists of a number of integrated modules in which the wave and flow fields, sediment transport and bed-level changes are computed sequentially. The morphodynamic process now is a high level of modelling sophistication, in which sediment transport and bed-level changes are an integral part of the flow module. In this study, the sediment version of Delft3D-FLOW is applied to carry out morphodynamic simulations, which can make alternating calls to the WAVE and FLOW modules. Fig. 2.4 shows the working procedure of the morphodynamic process.

In the morphodynamic process, a call to the Delft3D-WAVE module will result in a communication file being stored which contains the results of the wave simulation (RMS wave height, peak spectral period, wave direction, mass fluxes, *etc.*) on the same computational grid as is used by the FLOW module. The FLOW module can then read the wave results and include them in flow calculations. In situations where the water level, bathymetry, or flow velocity field change significantly during a FLOW simulation, it is often desirable to call the WAVE module more than once. The computed wave field can thereby be updated accounting for the changing water depths and flow velocities. At each call to the WAVE module the latest bed elevations, water

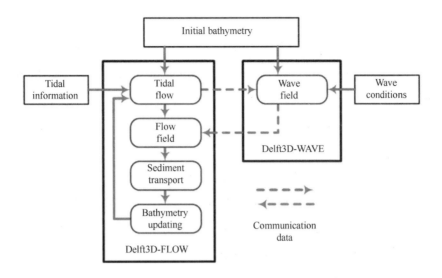

Fig. 2.4 Diagram of the morphodynamic process in Delft3D-FLOW

elevations and, if desired, current velocities are transferred from FLOW. The module is able to manage the simulation time for FLOW module and control the updating interval of wave computation. These functionalities provide conveniences to perform complex hydro and morphodynamic simulations.

In order to update the bed level, the exchange of sediment in suspension from the bottom computational layer to the bed (and vice versa) is modelled by means of sediment fluxes applied to the bed of each computational cell as

$$w_s c - \varepsilon_z \frac{\partial c}{\partial z} = D - E \qquad (2.13)$$

$$\Delta S_s^{(m,n)} = f_{MOR}(D - E)\Delta t \qquad (2.14)$$

where,

w_s ——sediment settle velocity (m/s);

c ——sediment concentration (kg/m^3);

ε_z ——vertical eddy viscosity (m^2/s);

D ——sediment deposition rate (kg/m^{2}s);

E ——sediment erosion rate (kg/m^{2}s);

$\Delta S_s^{(m,n)}$ ——net sediment change due to suspended load transport (kg/m^{2}s);

(m,n) ——computational cell location (-);

f_{MOR} ——morphological acceleration factor (–);

Δt ——computational (half) time step (s).

Delft3D computations are done on a staggered grid where the depth points are defined at the centre of each grid cell, velocity points at the mid-points of the grid cell side, and water level (and wave parameter) point at each grid cell corner. Sediment transport components, and sediment sources and sinks, are computed at water level point at each computational time step Δt. The bed level changes in the depth points are computed from the sediment transport gradients in the adjacent water level point that has the greatest water depth; this reduced shallow water numerical instabilities. The resulting change in the bottom sediment in each grid cell is added to the change due to the suspended sediment sources and sinks and included in the bottom updating scheme, thereby that the hydrodynamics are always calculated with the correct bathymetry. The bottom is updated at each computational time step.

Morphological changes take place on a time scale several times longer than typical flow changes; in the online sediment version of Delft3D-FLOW, the morphological acceleration factor f_{MOR} is used to deal with the difference in time scale between hydrodynamic and morphological development. It can be simply expressed as:

$$t_{\text{morphology}} = f_{MOR} t_{\text{hydrodynamic}} \qquad (2.15)$$

It thereby effectively extends the morphological time step by allowing accelerated bed-level changes to be incorporated dynamically into the hydrodynamic flow calculations. The introduction of f_{MOR} significantly reduces computational time, however, the maximum suitable f_{MOR} that does not affect the accuracy of the model, remains a matter of sensitivity testing for the individual situation (Grunnet et al., 2004).

2.5 Implementations of Delft3D system in present study

In this study, area model is applied to perform most computations. In addition, profile model is also used to calibrate some model factors, since the profile model can run with much less time efforts. Profile models in this study are all based on Delft3D system, and they can be thought as miniatures of area model (one cross section of area model). They almost share the same boundary conditions, initial conditions, physical parameters, and numerical parameters as the corresponding area models. In the following chapters, such model settings of area and profile models will be described in detail.

In this study, wave simulation can be performed by the Delft3D-WAVE module (SWAN) or SZW (Snell's and roller) or SWAN combined with roller, which potentially causes some confusion. So it is necessary to summary these different approaches. The first kind of approach, SWAN, is used in wave computation of hydrodynamic modelling, Chapter 4. The second approach SZW is used by profile modelling in Chapter 5. Combined SWAN and roller approach is also in Chapter 5, while the approach is adopted by area morphodynamic modelling. The detailed model setup of these approaches will be further discussed in the related chapters.

Except for the features mentioned above, the area model and the profile model may appear in different dimensions in this study. Comparison of the 2DH and 3D simulations of Delft3D (Lesser *et al.*, 2003) shows that the gradients in bed-shear stress are significantly reduced in the 3D simulation due to deformation of the logarithmic velocity profile. This smoother distribution of bed-shear stress is expected to cause the smoother development of the bathymetry in the 3D simulation.

For the area model, it can run in 2DH or 3D mode. Following the definition of the σ-coordinate, the former mode has only one layer while the latter one has multi-layers. Corresponding to the different dimensions of area models, the profile models also vary in dimensions. The profile model of a 2DH area model is in 1DH mode and that of a 3D area model is in 2DV mode. Table 2.5 summarises these model characteristics, and more details are also discussed in the following chapters incorporating with the related model setup.

Table 2.5 Model types of Delft3D system in present study

Model type	Flow grid dimension	Wave routine
Area	2DH (1 layer)	SWAN (Chapter 4)
	2DH (1 layer)	SWAN + roller (Chapter 5)
	3D (multi-layers)	SWAN + roller (Chapter 5)
Profile	1DH (1 layer)	SZW (Chapter 5)
	SZW (Chapter 5)	2DV (multi-layers)

Chapter 3

Egmond Shoreface Nourishment

In this chapter, the site of interest Egmond ann Zee and the shoreface nourishment are first introduced. Based on the data analysis which has been done in a previous study (van Duin *et al.*, 2004), the morphological aspects of the nourishment are further described.

3.1 Egmond beach

In 1990, a coastal policy was adopted to maintain the Dutch coastline position by applying shore nourishments. At erosive stretches, these nourishments have to be applied regularly every five years. Egmond aan Zee, a small seaside village along the North Holland coast, is such an erosion "hotspot". In the summer of 1999, a substantial shoreface nourishment volume was applied, backed with a beach nourishment volume. The main design objective of the shoreface nourishment is to improve coastline stability (i.e., to prevent the coastline from retreating landward). The shoreface nourishment is expected to diffuse in the cross-shore and longshore directions (Hamm *et al.*, 2002) and dissolve in the coastal system.

Egmond is located in the central part of the Dutch coast, between Den Helder and Hoek van Holland. Fig. 3.1 presents the geographical location of the site. The Dutch coast faces the North Sea and is exposed to sea waves and swell. The tidal wave, which finds its origin on the Atlantic Ocean, enters the basin of the North Sea in the north. The Coriolis force causes the tidal wave to rotate anti-clockwise in the tidal basin. Gradients both in phase and in amplitude occur along the Dutch coast. At Egmond, the general coast line orientation is 8°N (topographic North), which results in 278°N for the shore normal direction. The mean tidal range varies between 1.2 m in the neap tides to 2.1 m in spring tides. The tidal peak currents in the offshore zone are about 0.5 m/s; the flood current to the north is slightly larger than the ebb current to the south. The mean monthly offshore wave height has a seasonal character and varies from about 1 m in the summer months (May to August) to about 1.5 ~ 1.7 m in the autumn and winter (October to the next January). It may be as high as 5 m at 15 m depth during major storms from southwest or northwest directions. The beach width is about 100 m to 125 m with a slope between 1/30 and 1/50.

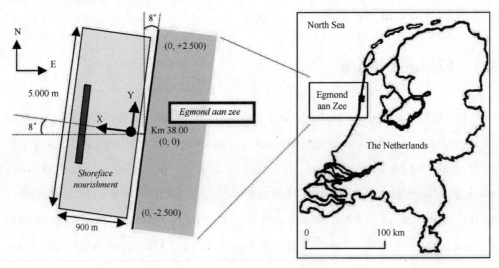

Fig. 3.1 Location of the shoreface nourishment at Egmond aan Zee. Left plot shows the plan view of survey area and shoreface nourishment area ($x=0$ represents beach pole line).

There are no hydraulic structures in the vicinity of the Egmond beach. The large-scale bathymetry can be characterised as an uniform and straight coast with parallel depth contours, dominated by waves. The small-scale morphology shows irregularities in the large-scale uniform pattern. This part of the Dutch coast is typical for the quasi-uniform sandy beaches dominated by breaker bars. Rip channels interrupt breaker bars and small, local bars are present. The coastal profile is a three-bar system: two breaker bars (inner and outer bars) in the surf zone and a swash bar (see Fig. 3.2). Two main longshore breaker bars run parallel to the shoreline most of the time. Fig. 3.3 shows a time-exposure video image of the Egmond beach (obtained by an Argus station, https://www.deltares.nl/en/projects/argus-video-systems/), in which the white strips indicate wave breaking on the surf zone bars. The shoreface nourishment was placed at the seaward flank of the outer breaker bar. The beach nourishment was placed at the beach behind the shoreface nourishment.

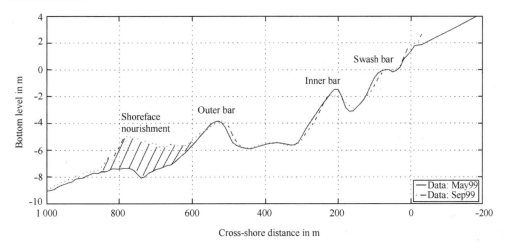

Fig. 3.2 Typical cross-shore profile at Egmond aan Zee with the shoreface nourishment

The inner bar is located at 200 m from the shoreline at 2 m below mean sea level, whilst the crest of the outer bar is located at about 500 m from the shore at 4 m below mean sea level. The inner bar is separated from the outer by a wide trough. Generally the area is characterised by medium well-sorted sands (0.25 mm ~ 0.5 mm), but in the trough between the inner and outer bars, sand is coarse (> 0.5 mm) and has moderate sorting. The cross-shore slope amounts to 1:100 and the median grain size is about 0.2 mm (Elias et al., 2000; van Rijn et al., 2001).

On large longshore scale (10 km) and on long term (years), the behaviour of the outer and inner bars at Egmond is two-dimensional movement in the sense that the bars are continuous and of the same form in

Fig. 3.3 Wave dissipation map based on time-exposure image

longshore direction and show the same overall migration pattern (onshore and offshore directions). On small scale (1km) and on the short time scale of a storm month, longshore non-uniformities may develop as local disturbances that are superimposed on the overall straight base pattern yielding a three-dimensional morphological system. Rip channels (with length of 200 to 300 m and depth of 0.5 to 1 m) are generated in the crest zone of the inner bar on the time-scale of a few days during minor storm conditions. Rip channels generally are washed out during major storm conditions. Overall, it can be concluded that the net changes at the inner bar and at the beach are relatively small, but larger changes can be observed at the outer bar. The bars show a long-term migration of about 20 to 40 m/year in seaward direction (van Rijn et al., 2003). According to Wijnberg (2002), the bars in the Egmond region show a periodic behaviour: the bars move offshore and the outer bar vanishes offshore while a new bar is generated near the shoreline.

In summary, the Egmond site shows the following key features:

(1) The coastal profile at Egmond is a three-bar system: two breaker bars in the surf zone and a swash bar;

(2) The outer bar is most pronounced, with its crest located at depths below NAP -3 m. The NAP (Normaal Amsterdams Peil) is the reference level to which level measurements in the Netherlands are related. A level of NAP 0 m is approximately equal to a mean sea level;

(3) The trough between the outer and inner bar is about 100 m wide and reaches the depths of NAP -5 m; the inner bar crest is located 300 m landward from the outer bar crest;

(4) Between the inner bar and the swash bar is a trough, which is less pronounced than the offshore trough and reaches depths of NAP -2 m;

(5) The bars and coastline show a rhythmical alongshore variation. With a longshore interval of approximately 2 km, the bars and coastline show a movement in onshore and offshore direction.

3.2 Egmond Shoreface nourishment

At Egmond aan Zee a shoreface nourishment has been applied in 1999. The center of the shoreface nourishment is in front of the lighthouse, Jan van Speijk (the RD coordinates 103011, 514782), or beach pole 38.00 km, see Fig. 3.1. The nourishment is approximately 2 kilometres long and 200 meters wide. The total sand volume is 900 000 m^3 with the characteristic volume 400 m^3/m. Fig. 3.4 shows the bathymetries before and after the implementation of the nourishment. The pre-nourishment bathymetry uses the measured data of May 1999, and the post-nourishment uses that of September 1999.

As part of the nourishment project, a comprehensive monitoring program of the bathymetry was setup. In Fig. 3.1, a plan view of the survey area (local grid system) is shown. The bathymetric data cover an alongshore length of 5 km and a cross-shore length of 900 m. The origin of the local coordinate system (0, 0) is situated at beach pole 38 and crosses the center of the shoreface nourishment area. There are eight times bathy data totally from 1999 to 2002. The nourishments applied during the investigated period are shown in Table 3.1.

Table 3.1 Dates of the available bathymetry data and data on the beach and shoreface nourishments

Bathymetry Date	Nourishments
May/June 1999	First beach nourishment: South boundary: RSP 38.750 (y = -750 m), North boundary: RSP 37.250 (y = +750 m) Characteristic volume of nourishment: 200 m^3/m; Total sand volume: 300 000 m^3
September 1999	Shoreface nourishment: South boundary: RSP 39.124 (-1 125 m); North boundary: RSP 36.875 (+1 124 m) Characteristic volume of nourishment: 400 m^3/m; Total sand volume: 900 000 m^3
May 2000	-
September 2000	Second beach nourishment: South boundary: RSP 38.800 (-800 m); North boundary: RSP 38.000 (0 m) Characteristic volume of nourishment: 258 m^3/m; Total sand volume: 207 000 m^3
April 2001	-
June 2001	-
October 2001	-
April 2002	-

The bathymetry data were obtained by both ship-based echo sounding and by the WESP (Water End Strand Profiler), using DGPS-based sounding (van Duin *et al.*, 2004). The WESP is a motorized three-legged vehicle. The shoreface nourishment, applied at Egmond aan Zee in the summer of 1999 (July and August), is approximately 2 km long and 200 m wide. The center of the nourishment area is in front of the lighthouse, Jan van Speijk. The other shoreface nourishment was carried out one year later, in September 2000.

Chapter 3 Egmond Shoreface Nourishment

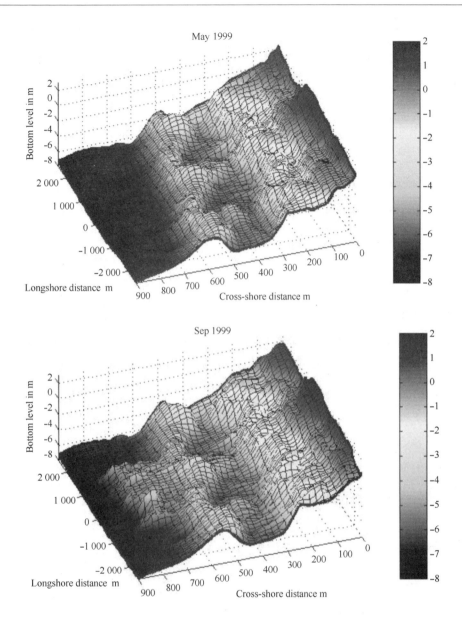

Fig. 3. 4 Egmond beach bathymetries pre- and post-nourishment

Based on the literature study, the following effects are expected to occur as a consequence of the placement of the Egmond shoreface nourishment (herein referred to as hypotheses, see Fig. 3. 4):

(1) Longshore effect: large waves break at the shoreface nourishment causing a calmer wave climate behind the shoreface nourishment area (wave filter) and a reduction of the longshore current and, hence, the transport

capacity. The shoreface nourishment acts as a blockade, resulting in: ① a decrease of the longshore transport; ② updrift sedimentation; ③ Downdrift erosion.

(2) Cross-shore effect: large waves break at the seaward side of the shoreface nourishment; remaining shoaling waves generate onshore transport due to wave asymmetry over the nourishment area; the smaller waves in the lee side generate less stirring of the sediment and the wave-induced return flow (cross-shore currents) reduces. This results in: ① an increase of the onshore sediment transport; ② a reduction of the offshore sediment transport.

Both effects result in an enhanced onshore transport behind the shoreface nourishment area. The objective of this book is to analyse and evaluate the morphological behaviour of the shoreface nourishment by analysing measured data and comparing these with model results. First, information is presented on the morphological changes based on regular soundings of the bathymetry. These cover the spatial scale of the surf zone (0.9 km cross-shore by 5.0 km longshore) and a time scale of several seasons (September 1999 to April 2002). Second, the model results generated with both a process-based profile model and a process-based coastal area model are analysed with the aim of testing the above mentioned hypotheses.

3.3 Morphological evolution analysis

3.3.1 Measured bathymetry development

Three-dimensional plots of the post-nourishment bathymetric data in 2000 and in 2001 are presented in Fig. 3.5 and Fig. 3.6, showing a shoreward migration of the outer bar and the formation of a trough between the outer bar and shoreface nourishment. The nourishment starts to behave like an outer bar and the original bar is forced to migrate onshore. The outer bar migrates onshore filling up the trough between the outer and the inner nearshore bar. During this migration, the original inner bar reduces or disappears. The radical change behind the nourishment is an indication of lee effect.

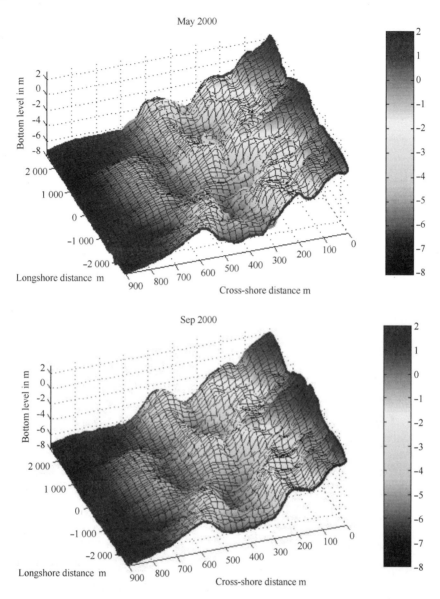

Fig. 3.5 Egmond bathymetries of May and September 2000

The shoreface nourishment seemed to act as the new outer bar and hardly changed in height and location (van Duin *et al.*, 2004). Therefore, it has not increased the beach sand volume directly, i.e., by redistribution of the nourished sand. In the same period, the inner bar also migrated shoreward. Both the outer and inner bars transformed into a boomerang shape (plan-

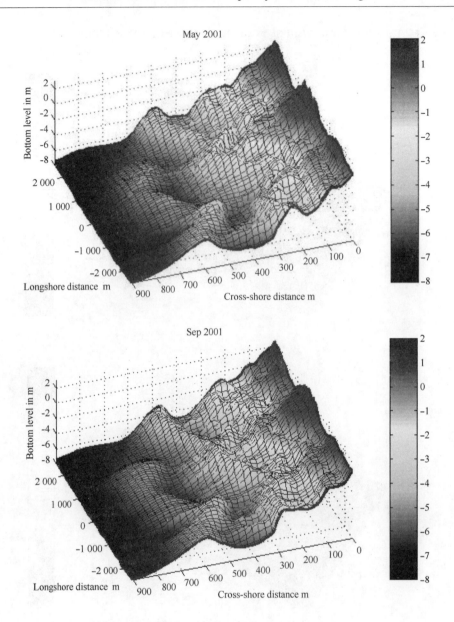

Fig. 3. 6 Egmond bathymetries of April and June 2001

form), which is also observed to some extent for the shoreface nourishment.

The survey in April 2002 (see Fig. 3. 7), however showed no further shoreward migration of the inner and outer bars. The outer bar had completely straightened and formed a continuous bar again. The shoreface nourishment decreased in height and lost its reef effect. It seems that the system was returning to its natural situation of a three-bar system: outer bar, inner bar and swash bar.

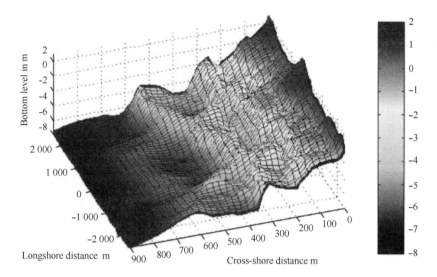

Fig. 3.7 Egmond bathymetry of April 2002

The basic assumption underlying the design and implementation of the shoreface nourishment is that eventually sand will be carried to the shore. The shoreface nourishment was expected to diffuse in the cross-shore and longshore directions, but the surveyed data showed that the shoreface nourishment did not diffuse much in the first two years. The scale (amplitude and length) of the shoreface nourishment was probably too large, as a result of which it did not diffuse and morph dynamic interaction occurred with the autonomous system (Hamm et al., 2002). After a period of two years, the shoreface nourishment started to diffuse and the bar amplitude at the shoreface nourishment area had become relatively small.

As is concluded in the data analysis of a long-term period (two years) from May 1999 to June 2001 (van Duin et al., 2002), the shoreface nourishment hardly changes in height and location. Therefore it has not increased the beach sand volume directly, i.e., by redistribution of the nourished sand. The investigated area shows a net gain of sand in the surveyed period, which has to come from longshore or cross-shore transport. An explanation for the stability of the nourishment can be the location. The nourishment was done in an area where morphological changes occur at a large time scale (the overall offshore bar migration cycle). The time scale in which morphological changes occur are years for the location of the shoreface nourishment. The beach zone, however, has a time scale of days in which large morphological changes can occur.

3.3.2 Volume change

To understand the observed morphological changes and derive a sediment balance, the coastal area of Egmond was divided into 20 subsections. The main criterion for the choice of the shore-parallel boundaries was to keep the (moving) bars in one subsection. Shore perpendicular boundaries were located at stable cross sections. Therefore, the shore-parallel boundaries were chosen at the troughs between the bars at $x = 600$ m and $x = 300$ m. The shoreface nourishment remains seaward of the shore-parallel boundary of $x = 600$ m. The nourishment area is split up into three subsections, containing a central part, a northern part, and a southern part. The subsections location is given in Fig. 3.8.

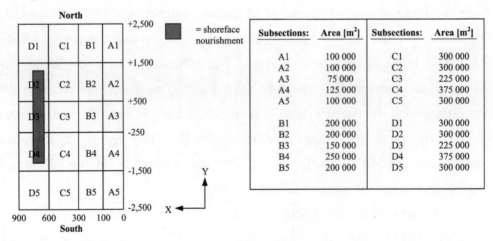

Fig. 3.8 Location of the subsections and their corresponding surface area

The subsections including the shoreface nourishment (subsections D2, D3, and D4) show a large volume increase caused by the construction of the shoreface nourishment during the period from May 1999 to September 1999. Over the total studied period, these three shoreface subsections show a volume increase of 550 000 m^3, including the placement of the shoreface nourishment, which means that 60% of the applied sand is still remained in three years. The second beach nourishment of July 2000 can be seen by the volume increase in the period of May 2000 to September 2000 in the beach subsections A3 and A4.

The shore-perpendicular subsections of the shoreface nourishment, i. e. subsections 2 to 4, all show a net increase of sediment for the period of May 1999 to April 2002. The total net gain of these three shore perpendicular subsections, including the placement of the shoreface nourishment and beach nourishment, is 747 500 m^3. The total sand volume of the shoreface nourishment is 900 000 m^3 and of the beach nourishment 207 000 m^3, which means that 65% of the applied sand is still present.

The sand volume change of the total area shows an increase for the first period, from May 1999 to September 1999, which is due to the placement of the shoreface nourishment. This caused an average bed level rise of approximately 0.16 m, a sand volume increase of 710 000 m^3. The total volume of the shoreface nourishment is 900 000 m^3. A volume of 190 000 m^3 is not recovered. Possible explanations can be measurement errors, truncation errors, transport of sediment out of the area, *etc.*. The volume increase continued until September 2000 and caused an extra average bed level rise of about 0.05 m, a sand volume increase of 240 000 m^3. For the period of September 2000 to April 2002, the average bed level shows a decrease. The total decrease is about 0.10 m and corresponds to a volume decrease of 470 000 m^3. In total, the area has lost a net sand volume of 230 000 m^3 after placement of the shoreface nourishment (September 1999 to April 2002), but a net gain of 477 500 m^3 for the overall period (May 1999 to April 2002) including placement of the shoreface nourishment. The total sand volume of the shoreface nourishment is 900 000 m^3 and of the beach nourishment 207 000 m^3, which means that 45% of the applied sand is still present after 3 years.

3.3.3 Longshore averaging of cross-shore profiles

In the study area, bars have an oblique orientation to the coast, and local depressions are present in bar crests and beaches. To identify the specific cross shore behaviour of the bar morphology, the bed profiles have been averaged over a sufficiently long alongshore distance (longshore scale). By this longshore averaging of cross-shore profiles, the variations due to alongshore phenomena such as sand wave patterns and rip channel patterns are eliminated. These phenomena will be expressed in a variation band (standard error) around the alongshore-averaged profile. The main idea behind this approach is to characterize each different longshore subarea (e. g. , subsections A3 to D3) by one cross-shore profile. This is only possible if the variation of the cross-shore profiles within an area is not too large, which is represented by the variation band.

Profile averaging has been performed for each survey by using transects (profiles) with a spacing of 100 m. These transects are averaged in shore-parallel direction. The study area is divided into five shore perpendicular subsections as mentioned in above Fig. 3. 8, which results in five longshore-averaged cross-shore profiles. These five longshore-averaged cross-shore profiles give an overall idea of the changes over time in cross-shore direction per shore-parallel subsection. The applied time intervals correspond roughly to either summer (spring and summer) or winter (autumn and winter). Therefore, seasonal effects can be detected easily.

The longshore-averaged profiles 1 (A1 to D1), 3 (A1 to D3), and 5 (A5 to D5), and the accompanying standard error, are shown in Fig. 3. 9. The plots show the longshore averaged profiles of May 1999, September 1999, April 2001, and April 2002. The plots of the longshore-averaged cross-shore profiles show a clear effect of the seasons. During the summer period, when relative little storms and storm surge levels occur, the morphological changes are small, whereas in winter large morphological changes occur.

In Fig. 3. 9b (the central part profile of the shoreface nourishment area), the shoreface nourishment can be seen on the seaward side of the outer bar at x = 650 m. Large morphological changes occur in the three subsections containing the shoreface nourishment. The beach volume decreases (shore-

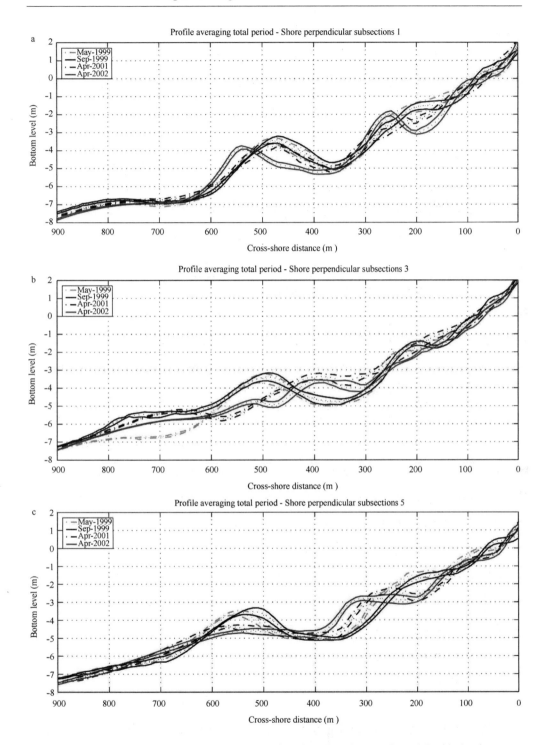

Fig. 3.9 Longshore averaged cross-shore profiles for subsections 1, 3, and 5 with the time of May 1999, April 2001, and April 2002.

parallel subsection A3), resulting in a lowering of the beach. The inner bar has moved slightly shoreward. The outer bar has moved about 150 m shoreward and is slightly flattened. In addition, the shoreface nourishment formed a flatter bar. After a period of 3 years, the system seemed to return to a more or less three bar system again. A more or less flattening of the profile can be seen. The shoreface nourishment and outer bar have decreased in height and the trough between the outer bar and shoreface nourishment has disappeared. The inner and outer troughs have deepened. Over the total period of 3 years, the shoreface nourishment subsections all show a volume increase, 110 ~ 390 m^3/m (the time interval from May 1999 to April 2002).

The area north of the shoreface nourishment (shore-perpendicular subsections 1) shows smaller morphological changes (see Fig. 3.9a). Over the total period, both the outer and inner bars moved seaward slightly. The inner bar is more pronounced, while the outer trough has widened. The beach zone stayed more or less the same over the total 3-year period. Overall, there is a small volume decrease of about 150 m^3/m.

For the area south of the shoreface nourishment (shore-perpendicular subsections 5), a clear flattening of the profile can be seen (see Fig. 3.9c). The averaged profiles shows a strong lowering of the outer bar (about 1.5 m), which has almost disappeared. The inner bar height is smaller and has moved approximately 50 m seaward. The beach shows an overall loss of volume of 120 m^3/m. The northern and southern sections (profiles 1 and 5) both show a relatively large volume decrease over 3 years, which is much larger than the natural autonomous erosion.

3.4 Conclusions

The bathymetry data of Egmond aan Zee, covering a period of May 1999 to April 2002, have been analyzed to study the influence of the shoreface nourishment area on the morphodynamics. During the first 2 years of the study period the shoreface nourishment hardly changed in height or location and, therefore, did not contribute to the beach sand volume directly, i.e., by redistribution of the nourished sand.

The inner and outer bars, on the other hand, showed a large shoreward migration and a trough was generated between the outer bar and the shoreface nourishment area. The shoreface nourishment area seemed to act as the new outer bar, taking over the function of the original outer bar. The survey of April 2002 showed an end to this trend. The system seemed to return to its natural three-bar system.

The shoreface nourishment was expected to diffuse, but surveyed data showed that the shoreface nourishment did not diffuse in the first two years. The scale (amplitude and length) of the shoreface nourishment was probably too large as a result of which it did not diffuse and morphodynamic interaction occurred with the autonomous system. After a period of 2 years, the shoreface nourishment started to diffuse resulting in a lower amplitude and shoreward movement of the outer bar.

Accretion occurred shoreward of the shoreface nourishment, indicating that the shoreface nourishment functions as a reef with a lee-side effect shoreward of the nourishment area. The total area shows a net gain of sand volume during the overall period (May 1999 to April 2002), including placement of the shoreface nourishment and beach nourishment. After three years, about 45% of the nourishment is still present.

Chapter 4

Hydrodynamic Modelling

The present book consists of a hindcast study to validate the modelling program Delft3D on the Egmond shoreface nourishment. This chapter describes the setup of the Egmond Delft3D model. The hydrodynamic validation (flow and wave) of the model is carried out, which is the basis of the morphodynamic simulation discussed in the next chapter. Transport is an important topic related to hydrodynamic validation, which is computed based on the local flow and wave conditions. Sediment transport links hydrodynamic conditions and morphodynamic responses. This issue is described in the last section of this chapter, but it is also discussed in the next chapter together with morphodynamic validation.

In this chapter, Section 4.2 describes the design of the computational grids for flow and wave modules, and the bottoms related to the grids. In Section 4.3 the tidal schematisation is presented, after which in Section 4.4 the wave schematisation is described. Wind-induced flow is not included in the model due to the lack of data. The corresponding parameters set to the flow module are also introduced in Section 4.3. Then, the calibrations of tidal current are carried out against to Egm2002. Section 4.4 discusses the hydrodynamic computations which consist of all the schematised wave conditions. Further in Section 4.5, the hydrodynamic results of wave-current interactions are analysed. Finally, the sediment transport is discussed in Section4.6.

Chapter 4　Hydrodynamic Modelling

4.1　Computational grids and bathymetries

In this book, a new Delft3D model is built which is different from the previous 2DH model (van Duin, 2002). To avoid confusing these two models, the previous model is called Egm2002 and the new model is named Egm2004. The new model Egm2004 appears in 2DH (1 layer) and 3D (11 layers) modes in the study. Egm2004 is calibrated against Egm2002 (2DH), so it is in 2DH mode during hydrodynamic modelling. Finally the model will be expended to 3D mode to perform morphodynamic validations, which is described in the next chapter.

Before starting the simulation, computational grids have to be made. Using the measured depth values, the bathymetry can be interpolated to the grids. The wave calculations are performed on a wave grid, which in this case is larger than the flow computational grid. The flow or morphology grid is nested within the wave grid. The flow grid has to be large enough to keep boundary disturbances out of the area of interest, and the boundaries of wave grid also should be far enough to prevent the wave boundary disturbances from the morphology grid.

Both grids (wave and flow) used in the study are curvilinear grids which are designed with the program RGFGRID, one of the Delft3D modules. The RGFGRID program is used to create, modify and visualise model grids for the other Delft3D modules. Another Delft3D module QUICKIN is used to interpolate the depth values to the computational grids. The QUICKIN program is a powerful tool to create, manipulate and visualise model bathymetries for the Delft3D computational modules. For detailed information about these tools, see the corresponding manuals of the software modules published by Delft Hydraulics (Deltares).

4.1.1　Computational grids

The computational grids are chosen longshore, therefore having a clockwise rotation of eight degree in relation to the true north. The flow grid is nested within and overlapped to the wave grid. Fig. 4.1 shows both grids. The grids are roughly centred on the Jan van Sperk lighthouse in longshore direction. They have various resolutions on space. The more close to the area of interest (nourishment), the higher the resolution of flow grid becomes. In the flow grid, the size of grid cells is about 20 m in cross-shore direction and 35 m along longshore direction around the shoreface nourishment.

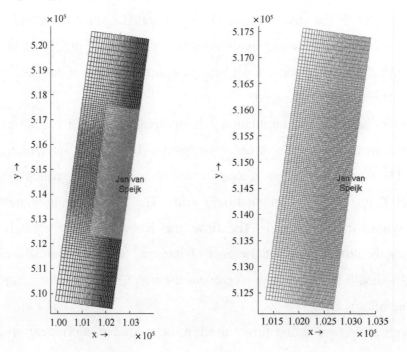

Fig. 4.1 Wave and flow grids of Egm2004. Left plot shows the wave grid (deep grey) and the flow gird that is amplified in the right plot.

On the boundaries it is about 65 m cross-shore and 90 m longshore. The total longshore grid length is approximately 5 200 m. The total cross-shore grid width is about 1 300 m. It has a total of 112 grid cells in longshore direction and 48 cells in cross-shore direction. The total number of grid cells is up to 5 376.

In the Egm2004 model, boundary-fitted σ-coordinates are used in the vertical direction. The vertical grid consists of layers bounded by two σ-plane, see Fig. 2.2. This means that over the entire computational area, irrespective of the local water depth, the number of layers is constant. As a result, a smooth representation of the topography is obtained. The relative layer thickness are usually non-uniformly distributed, which allows for more resolution in the zones of interest, such as the near surface area (important for e. g. wind-driven flows, breaking waves, heat exchanges with the atmosphere) and the near bed area (sediment transport). In the Egm2004 model, the vertical profile of the flow grid is specified to 11 layers, shown in Fig. 2. 3. Since the model aims to simulate morphological changes, the thickness of the bottom layer should be small. Wave dissipation is also an important issue in the study, so the surface layer should be small. The variation in the layer thickness should not be larger, i. e., the layer thickness must have a smooth distribution. An indicative value for the variation-factor for each layer is from 0. 7 to 1. 4. In Egm2004, the vertical profile has 11 layers with the thicknesses of 2 – 5 – 8 – 10 – 15 – 20 – 15 – 10 – 8 – 5 – 2% respectively.

In order to minimize truncation errors in the finite difference scheme which is used in Delft3D, the grid should satisfy the requirements of orthogonality and smoothness. According to the Delft3D manual, the error in the computed direction of the pressure term is proportional to the orthogonality values, so these values should be smaller than 0. 05 (preferably <0. 02). In the grids of Egm2004, these values stay under 0. 01, which is sufficient to have a negligibly effect on the exactness. Adjacent grid cells should vary less than 20%, although local exceptions may be acceptable. For the Egm2004 model the M-smoothness (cross-shore) is less than 7%. The overall N-smoothness (longshore) is also below 7%. All the characteristic values of the Egm2004 grids can meet the requirements of the modelling system. The parameters of both girds (wave and flow) are listed comparatively in Table 4. 1.

Table 4.1 Properties of wave and flow grids

Property		Wave grid	Flow grid
Cross-shore	Distance	2 400 m	1 300 m
	Number of cells	59	48
	Size of cells	20 ~ 133 m	20 ~ 65 m
	Smoothness	< 1.08	< 1.07
	Curvature	< 0.79	< 0.13
Longshore	Distance	10 900 m	5 200 m
	Number of cells	150	112
	Size of cells	36 ~ 232 m	36 ~ 87 m
	Smoothness	< 1.08	< 1.07
	Curvature	< 0.24	< 0.24
Total grid cells		8 850	5376
Resolution		27 ~ 176 m	27 ~ 75 m
Orthogonality		< 0.01	< 0.01
Aspect-Ratio		< 11.58	< 4.31
Vertical resolution (layers)		1 (2DH)	1 (2DH) /11 (3D)

4.1.2 Model bottoms

The first available bathymetry data set after the shoreface nourishment is dated on 01-09-1999. The bathymetry samples consisting of the coordinates (x, y) and water depth (z) of the measured points are loaded into the computational grid in the QUICKIN program. By means of triangular interpolation these depth values are interpolated to the computational grid of the Egm 2004 model. The samples can fully cover the flow grid, but a few cells on the outer edges of the wave grid are lack of sample points. Using the line sweep method of QUICKIN, such area adopts the samples nearby. The bathymetries of the grids produced by QUICKIN are shown in Fig. 4.2. The bottom of the flow grid is isolated by a white line within the wave grid. In the flow grid, three cross sections (north, middle and south) are shown. The middle section is located in the middle of the nourishment, more or less the center of the flow grid. The north and the south sections are located to the north and south of the nourishment, respectively. These sections are used to evaluate the modelled results in the following chapters.

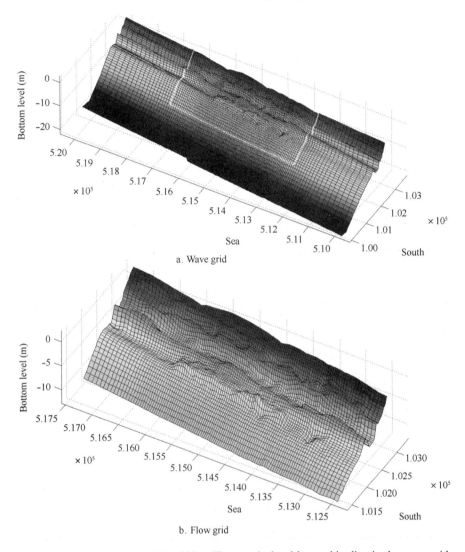

Fig. 4.2 Model bottoms of Egm2004. The area isolated by a white line in the wave grid is the flow grid. The three cross sections in the flow grid will be used to present modelled results in the following sections.

4.2 Tidal schematization and calibration

The schematized tide in this study is the same as used in Egm2002. Based

53

on the tide information, the boundary conditions are generated for the Egm2004 model. Egm2004 adopts most of physical and numerical parameter settings from Egm2002. Then Egm2004 is run in 2DH mode without wave coupled. The results (tide currents) are tested against Egm 2002.

4.2.1 Tidal schematization and boundary conditions

The tidal information is used for hydrodynamic computation and morphological simulation. Ideally one would like to simulate a complete tidal cycle (e.g. neap-spring tidal cycle), but this would lead to an unacceptable high computational effort. So to minimize the computational time a tidal schematization has to be made. The tidal schematization is to derive a representative tide (i.e. a morphological tidal cycle) which results in a reliable description of the net sediment transports in the study area has to be selected first.

In this study, the representative tide of the previous Egm2002 model continues to be used, since the Egm2002 model is used to calibrate the new Egm2004 model. In order to find the harmonic components of the representative morphological tide, which represents the tide representative sediment transport, a Fourier analysis was done in the above mentioned report. The harmonic components which result from the analysis, were used as hydrodynamic boundary conditions for the computations within the Egm2002 model. The advantage of harmonic components over time series as a hydrodynamic boundary conditions is the more efficient computation time. In Egm2002, the northern and seaward (western) boundary conditions are water levels (harmonic), the condition at the southern end is uniform velocities normal to the boundary.

In the present study, the seaward boundary condition is still harmonic water levels which reproduce the representative tide. The conditions at the northern and southern ends are controlled by Riemann invariant (weakly reflective boundaries). Instead of a fixed water level or velocity, the weakly reflective boundary condition is the longshore water level gradient. The main characteristic of a weakly reflective boundary condition is that the boundary up to a certain level is transparent for out-going waves, such as short wave disturbances. Out-going waves can cross the open boundary without being reflected back into the computational domain as happens for other types of boundaries. In many cases, the longshore gradient of the water level does not vary much in cross-shore direction, so an uniform boundary condition can be assigned at each lateral cross-shore boundary.

As mentioned above, Egm2004 and Egm2002 use the same representative tide. But the area of the morphological grid in Egm2004 is different from the previous model, the boundary conditions have to be regenerated. Fig. 4.3 gives the boundary locations of Egm2002 and Egm2004. In the right plot, the time-series water levels are plotted for the corresponding stations of the left plot. The tide range in the area of interest is about 1.6 m.

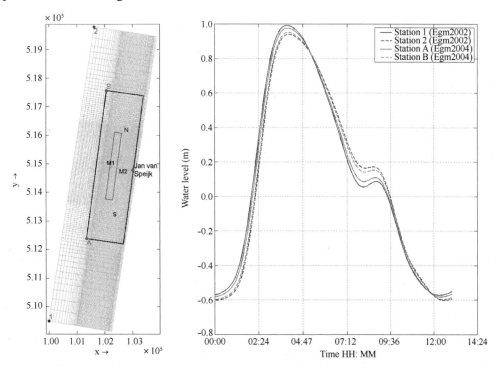

Fig. 4.3 Boundaries and the schematised morphological tide in Egm2004. The grey grid in the left plot is Egm2002. The black box is the area of Egm2004. The red box indicates the location of the nourishment. The observation stations S, N, M1, and M2 are used to compare the computed results of both models.

The propagation of the representative tidal water level along the coast can be described by:

$$\eta(s,t) = \sum_{j=1}^{N} \hat{\eta}_j \cos(\omega_j t - \kappa_j s - \varphi_j) \qquad (4.1)$$

where,

η ——water level (m);

$\hat{\eta}_j$ ——amplitude of the j-th harmonic component (m);

ω_j ——angular frequency of the component (°/h);

κ_j ——longshore wave-number of the tidal component (-);

s ——longshore distance (m);

φ_j ——phase lag of the component (°).

To obtain the longshore gradient of the water level we can now simply differentiate Equation (4.1) with respect to the longshore distance, then get:

$$\frac{\partial \eta}{\partial s}(s,t) = \sum_{j=1}^{N} k_j \hat{\eta}_j \sin(\omega_j t - \kappa_j s - \varphi_j) = \sum_{j=1}^{N} k_j \hat{\eta}_j \cos\left(\omega_j t - \kappa_j s - \varphi_j - \frac{\pi}{2}\right)$$
$$(4.2)$$

The model area of Egm2004 is in the middle of the Egm2002 model. Using the tidal information of the water level stations (1, 2) of Egm2002 where tidal amplitudes and phases are known, the water level amplitudes and phases of the stations A and B of the new grid can be determined by spatial interpolation according to Equation (4.1). The longshore wave-number can be derived for each component by analysing the phase difference between the two stations. Then using Equation (4.2), the longshore gradient of water level for each station can be derived. So at the seaward boundary the water level is prescribed, and at the lateral boundaries is a uniform longshore gradient of water level as a function of time with the combination of harmonic components. For the seaside boundary, Table 4.2 and Table 4.3 list the detailed characteristics of harmonic component of the corresponding stations in Fig. 4.2. The lateral boundary conditions, i.e. Riemann invariants which are derived from Table. 4.3, are summarized in Table. 4.4. For more details see (Roelvink and Walstra, 2004).

Chapter 4 Hydrodynamic Modelling

Table 4.2 Harmonic components of Egm2002

Angular velocity (°/hour)	Station 1		Station 2	
	Amplitude (cm)	Phase (°)	Amplitude (cm)	Phase (°)
0.0	13.260	0.000	12.789	0.000
28.8	70.360	146.779	70.571	155.752
57.6	26.576	-142.362	24.857	-137.581
86.4	4.063	-30.774	6.059	-26.633
115.2	7.262	-10.309	6.928	-1.783
144.0	0.346	128.204	1.514	141.743
172.8	1.649	124.719	1.866	142.243

Table 4.3 Harmonic components of Egm2004

Angular velocity (°/hour)	Station A		Station B	
	Amplitude (cm)	Phase (°)	Amplitude (cm)	Phase (°)
0.0	13.121	0.00	12.884	0.00
28.8	70.422	149.43	70.528	153.93
57.6	26.068	-140.95	25.205	-138.55
86.4	4.653	-29.55	5.654	-27.47
115.2	7.163	-7.79	6.996	-3.51
144.0	0.691	132.21	1.277	139.00
172.8	1.713	129.90	1.822	138.69

Table 4.4 Riemann invariants of Egm2004 lateral boundaries

Angular velocity (°/hour)	South		North	
	Amplitude (10^{-5} cm)	Phase (°)	Amplitude (10^{-5} cm)	Phase (°)
0.0	0.000	0.00	0.000	0.00
28.8	105.520	239.43	105.830	243.93
57.6	21.222	-509.49	19.849	-485.53
86.4	2.812	604.50	4.193	625.28
115.2	10.348	822.12	9.872	864.89
144.0	0.783	222.21	3.426	229.00
172.8	4.830	219.90	5.465	228.69

4.2.2 Computational parameter settings of hydrodynamics

Most model parameters use their default values. But some parameters specified to the site of interest are worth mentioning. Selections of time step and bottom roughness coefficients (Manning formula is used in this study) are described below.

The hydrodynamic module solves the unsteady shallow water equations (assuming that the vertical accelerations can be neglected) in two (depth-averaged mode) or three dimensions on a curvilinear grid system. These equations are solved with the Alternating Direction Implicit (ADI) method in the horizontal direction, and with a fully implicit scheme in the vertical direction (WL | Delft Hydraulics, 2003a). Since this solution is implicit, the numerical stability is not restricted by the time step Δt or the grid size. However since the accuracy of the flow decreases with the increase the time step, a widely used parameter for behaviour of the flow, the Courant number σ, is evaluated:

$$\sigma = 2c\Delta t \sqrt{\frac{1}{\Delta x^2} + \frac{1}{\Delta y^2}} \qquad (4.3)$$

$$c = \sqrt{gh} \qquad (4.4)$$

where,

σ ——Courant number (-);

c ——wave celerity (m/s);

Δt ——time step (s);

Δx ——grid dimension in x direction (m);

Δy ——grid dimension in y direction (m);

g ——acceleration due to gravity (m/s^2);

h ——local water depth (m).

The Courant number gives the relationship among the time step, the wave propagation celerity and the grid size. To obtain accuracy in Delft3D-FLOW, the Courant number is an indication. The number should exceed a critical value to ensure accuracy; a value of $4\sqrt{2}$ was suggested in the manual.

Chapter 4 Hydrodynamic Modelling

Meanwhile, the magnitude of the time step determines the total computation time. To reduce the total computational time, it is necessary to choose the largest time step possible, without loss of accuracy and stability. Several sensitive runs carried out under varying time steps showed that the flow results are identical using the time steps between 12 and 30 seconds.

The directives for the Courant number are based on experience. In practical situations the Courant number should be in the order of 10. This is however a rough estimate and sensitivity runs should be carried out in order to determine the maximum time step for which Delft3D still yields accurate results. The time step limitation is not only related to hydrodynamic performance, but also depends on morphological sensitivity. To concern the morphology updating, the time step larger than 12 seconds may cause computation unstable in this case. For the Egm2004 model, sensitive runs show the time step of 12 seconds satisfies the stability and accuracy of hydrodynamic computation as well as morphodynamic simulation. So the time step in the study is finally set to 12 seconds. But at this moment, the morphological factor settings in the sensitive runs are not considered systematically to agree with the final morphological scenario which is discussed in the next chapter. Since a longer time step in morphological simulation can significantly save computation time, to test the sensitivity of different time steps to the mor-phological computation stability and accuracy is highly recommended in the future study.

The bottom roughness formula used in the study is Manning's, both values of longshore and cross-shore coefficients are 0.026 $m^{1/3}/s$, which is equivalent to the Chezy coefficient $C_{2D} \approx 56$ $m^{1/2}/s$. This is based on the input values of the earlier Delft3D studies at Egmond aan Zee which has been done by Klein and Elias (2001), van Duin et al. (2004). The Coriolis effect is taken into account in the computation, which is related to the geographic latitude of the site of the interest. The water temperature is 8 ℃. The density of the water is 1023 kg/m^3. These parameters follows the setting in the previous studies mentioned above. Table 4.5 summaries these parameters of model settings.

Table 4.5 Parameter setting for Egm2004-FLOW module

Parameter	Value	Unit	Description
Δt	12	s	computational time step
n	0.026	$m^{1/3}/s$	bottom friction (Manning coefficient)
δ_H	1.0	m^2/s	horizontal eddy diffusivity
υ_H	10.0	m^2/s	horizontal eddy viscosity
Dry/Flood	mean	–	determination for drying/flooding in grid cell
H_{dry}	0.4	m	threshold depth for drying and flooding
g	9.81	m/s^2	gravity
ρ	1023	kg/m^3	water density
T	8	℃	water temperature
S	31	ppt	salinity

4.2.3 Calibration of tidal flow

The Egm2004 model is calibrated against the previous Egm2002 model. Except the different areas of both models, they have different types of boundary conditions and time steps. The identical representative tide is adopted by two models, so both models actually have the same driving forces. In the new model Egm2004, the hydrodynamic boundary conditions are given as Riemann condition in the north and south lateral ends, and harmonic water levels in the seaside. As explained in Section 4.3.1, in Egm2002 the conditions are harmonic water levels (north and sea boundaries) combined with uniform velocities (south side). Both models use the same physical and numerical parameters listed in Table 4.5, except for time steps. The time step in 2002 is 30s, while it is 12s in Egm2004. Two models use different approaches to deal with sediment transport and morphological change, which bring different time steps to satisfy the requirement of morphodynamic simulation. This issue will be further discussed in the next chapters.

The main checks are time series of water level and velocity. The calibration of water level is aiming at the accurate reproduction of tidal water levels. Fig. 4.4 shows the comparison of water levels and velocities of the station M1 (see Fig. 4.3) in both models. The station M1 is located on the outer slope of the nourishment. The station in two models is not an exactly identical point but within about 25 m distance, since both grids are not overlapped completely.

From the figure, both models have the same time series of water level in the station. The longshore velocities almost coincide to each other, but the obvious difference can be observed at low water. This phenomenon is possibly caused by interpolation errors in regeneration of boundary conditions, and the different types of boundary conditions. The maximum flood longshore velocity is about 0.5 m/s, and the maximum ebb is close to 0.4 m/s, slightly less than the flood velocity. The high water lags about 2 hours behind the maximum flood, and the low water does about 4 hours behind the maximum ebb. For cross-shore velocities, differences are somewhat larger, which maximum is about 0.05 m/s and happens during the slack water of ebb current. The velocities vary between ± 0.05 m/s in Egm2002 and ± 0.02 m/s in Egm2004. The differences are potentially owing to the different types of boundary conditions used in two models, the different grid resolutions, and the interpolated errors of bathymetries between both models.

The Riemann conditions can bring more smooth cross-shore velocities, and make the computation stable earlier than other boundary conditions, which can reduce the spin-up time. Fig. 4.4 shows time series starting from 12:30 and lasting 37.5 hours (3 tides). The first 12.5 hours is omitted from the figure, since this period belongs to the spin-up time. According to the lower plot in the figure, Egm2004 becomes stable after 15 hours from initial conditions, and Egm2002 needs at least 3 more hours. The comparison results at the station N, M2 and S (see the locations in Fig. 4.3) are shown in Appendix Fig. A.1. These results at different observation stations also show that excellent agreements are reached between both models, except for cross-shore velocities, but the values are small (<0.05 m/s).

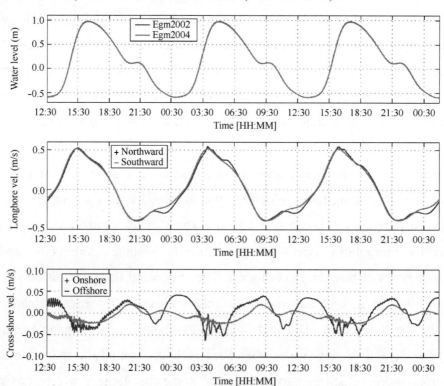

Fig. 4.4 Comparison of water levels, longshore and cross-shore velocities at Station M1

Fig. 4.5 shows the depth-averaged flow vectors at different time which is corresponding to the last tide cycle in Fig. 4.4. The figure indicates that Egm2004 well reproduces the flow field of Egm2002 (for clarity reasons, only a fraction of the grid points are presented here). At low water, there are larger onshore velocities happening in Egm2002. Meanwhile, we can see more uniform solutions under Riemann boundary conditions. Tide-averaged velocities modelled by both models are shown in Fig. 4.6. Both models produce northward tide-average velocities. In a total manner, the velocities in Egm2004 almost point to land in cross-shore sections, but the velocities in Egm2002 are somewhat disorderly. This situation might be explained by the discrepancies of the boundary conditions and the model areas.

Chapter 4 Hydrodynamic Modelling

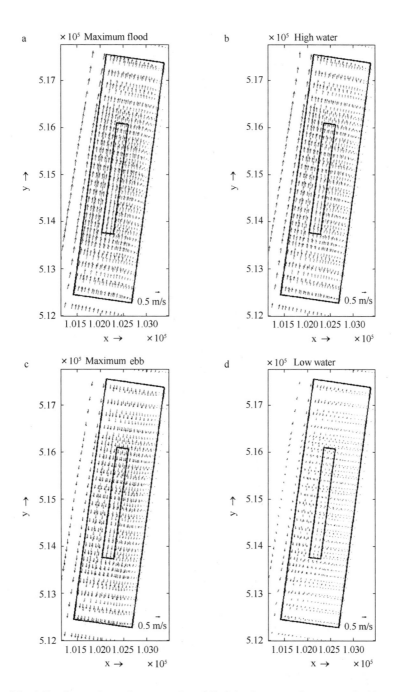

Fig. 4.5 Comparison of vectors of modelled depth-averaged current velocities.

a. Upper left plot, maximum flood 16:30, b. Upper right plot, high water 17:30, c. Lower left plot, maximum ebb 22:00, d. Lower right plot, low water 02:00. The location of the nourishment is indicated by the small box.

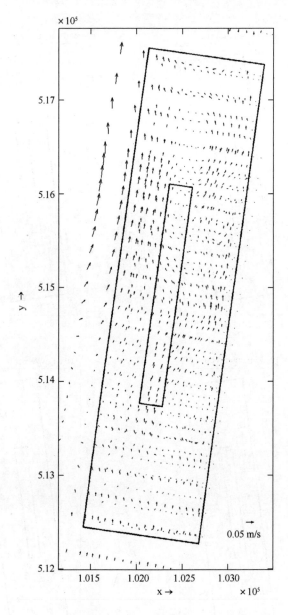

Fig. 4.6 Comparison of tide-averaged current velocities.
Egm2002 shows in blue arrows and Egm 2004 does in red ones.

4.3 Wave modelling

To perform wave simulations, the SWAN model of the Delft3D-WAVE module is used. Although SWAN model can fulfil wave computation independently, it is coupled with Delft3D-FLOW module in present simulations. Delft3D-FLOW steers this coupling. The Egm 2004 is still run in 2DH mode to carry out the wave modelling and take wave-current interactions into account. The wave boundary conditions are a set of wave conditions schematized from measured wave climate, which is first discussed in the following subsections. The modelled results are then presented as wave heights and energy dissipations.

4.3.1 Wave schematization

A wave schematization aims at reducing the wave climate into a choice of representative wave conditions. Since no wind data are available, the modelling performances are without wind. Similar to the tidal schematization, the wave schematization is also to derive morphological wave boundary conditions. This schematization should result in a reliable description of the net transports due to wave climate. In this section, the schematized wave conditions in the previous study (van Duin, 2002) are briefly introduced. But only the hydrodynamic character related to the schematized waves are discussed. The sediment transport induced by the waves will be described in the next section.

The schematization method is given in two steps. Step 1 is to make the division of the given wave time series in sectors and the choice of wave heights in combination with wave direction. Step 2 is to do the calculation of the required simulation time of the grouped waves, according to the net transport per wave condition.

To give a good representation of the actual wave developments, the existing offshore wave records are considered. The used wave climate is measured from IJmuiden for the period of September 1999 to May 2000. To derive a representative set of wave boundary conditions, the wave climate has to be classified in different wave heights and directions. Then these classes are grouped to 6 sectors by combining wave heights and directions. It was tried to make the amount of data points per sector more or less equal. Table 4.6 lists the probability of occurrence $P(i)$ for each sector with the total probability of about 15%. In order to single out the effect of storm events on nearshore bathymetry, two wave sub-sectors which correspond to an average wave height and a high wave height are further separated.

Table 4.6 Probability of occurrence of wave heights per direction in percent

H_S	Wave direction (average valve of considered class) in degrees																	
	185	195	205	215	225	232.5	237.5	245	255	265	275	285	295	305	315	325	335	345	335
0.25	0.75	0.43	0.77	0.63	0.91	0.39	0.41	0.68	0.63	0.50	0.34	0.61	0.91	1.50	2.16	2.66	3.14	4.02	1.13
0.75	0.30	1.30	1.98	2.70	3.05	0.78	0.73	1.34	1.43	1.20	1.23	1.50	1.61	2.82	3.14	3.77	4.11	1.79	0.71
1.25	0	0.09	0.93	4.23	4.43	1.16	0.84	0.91	0.75	0.95	0.84	0.71	1.00	1.57	1.43	1.54	1.05	0.41	0.75
1.75	0	0	0.13	1.16	1.77	0.82	0.41	0.48	0.52	0.54	0.41	0.71	0.77	0.43	0.82	1.50	0.23	0.45	0
2.25	0	0	0	0.05	0.91	0.66	0.30	0.50	0.25	0.27	0.23	0.41	0.11	0.14	0.46	0.50	0.18	0.18	0
2.75	0	0	0	0	0.41	0.11	0.05	0.05	0.02	0.04	0.14	0.21	0.21	0.11	0.27	0.13	0	0	0
3.25	0	0	0	0	0.02	0.16	0	0.04	0.02	0.05	0.07	0.09	0.09	0.02	0.05	0.02	0	0	0
sum			15.4			15.6			14.0				18.4			18.5			18.1

The UNIBEST-TC program is used to predict the net transports due to the input wave climate. (UNIBEST, a software developed by Delft Hydraulics, stands for UNIform BEach Sediment Transport. TC stands for Time dependent Cross-shore, but this model includes longshore option.) The model is executed without the bottom updating. Running these different wave conditions for their own occurrences in UNIBEST-TC leads to schematised net sediment transports. The summation of the sediment transports of the averaged wave condition and the high wave condition with the same direction, gives the total net schematised sediment transport per directional sector. The longshore sediment transport calculated by the model is used to derive the morphological time for each wave sector. This work has been done in the previous study (van Duin, 2002), and the results are adopted in this study. An assumption is made that if the sediment transport due to the schematized wave conditions is reliable in UNIBEST-TC, it also will do in Delft3D. The above mentioned schematization method finally results in twelve wave conditions (six directions times two wave heights) with simulation durations, which together result in an overall sand transport comparable to the sand transport if all the occurred wave conditions would be considered. Table 4.7 lists the schematised wave conditions, i.e., H_s significant wave height, T_s wave period, and $T_{morphology}$ schematised wave duration. Fig. 4.7 describes the wave roses pre- and post-schematisation.

Table 4.7 Schematised wave conditions and morphological time

Wave condition	Direction (°N)	H_s (m)	T_s (s)	$T_{morphology}$ (days)
1	205	0.75	5.0	96
2	205	1.65	7.0	6
3	225	1.25	6.3	46
4	225	2.75	8.3	4
5	245	1.25	6.3	11
6	245	2.25	7.8	6
7	295	1.25	6.5	19
8	295	2.75	8.5	4
9	325	1.25	7.5	31
10	325	2.75	9.5	3
11	345	0.75	5.6	35
12	345	2.25	8.7	3

4.3.2 Wave heights

Detailed calibration and validation of a wave model is not possible without time-series of measured wave data from the area of interest. In the present study default settings of the Delft3D-WAVE module (SWAN) have been used to carry out wave simulations.

Waves approaching the shore undergo a systematic transformation. In the offshore region, wave height decreases as a result of energy dissipation due to bottom friction. As waves propagate further shoreward, the wave celerity and wavelength decrease and the wave height increases (shoaling), leading to an increase of wave steepness, see Fig. 4.8. Waves approaching the shore under

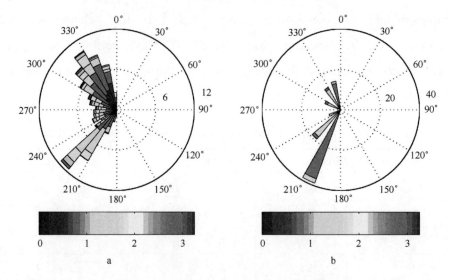

Fig. 4.7 Wave roses of pre-and post-schematisation.

a. Left plot, easured wave climate, b. Right plot, schematized wave conditions. The radium indicates the probability of occurrence of wave heights. The maximum radium of the left plot is 12%, while the right plot is 40%.

an angle gradually reorient (refraction), near the beach eventually leading to the wave crest moving parallel to the shoreline.

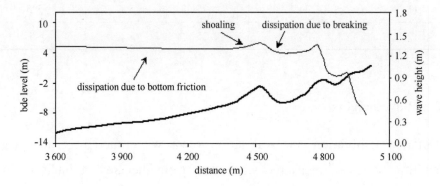

Fig. 4.8 Shoaling and breaking of waves across a nearshore profile with sandbars

The wave module of Egm2004 is driven by the schematised wave boundary conditions. The module accounts for refractive propagation over the bathymetry and currents which are computed by the flow module. Especially for oblique incident waves along the lateral boundaries, uniform wave heights are not in accordance with local water depths thereby introducing disturbances at these boundaries. So a larger wave grid is built to prevent such disturbances from the morphological grid, see Section 4.2. The bathymetry and the currents outside the flow grid are derived by interpolating the flow results in the communication file to the wave grid. In the wave simulation, the processes of whitecapping, frequency shift, bottom dissipation and depth-induced wave breaking are represented. Some parameters used in the module are summarised in Table 4.8.

Table 4.8　Parameter settings for Egm2004-WAVE module

Parameter	Value/Method	Description
Spectrum	JONSWAP	shape of the wave spectrum
γ (spectrum)	3.3	peak enhancement factor (JONSWAP)
Setup	false	wave-related water level setup
Forcing	energy dissipation	computation of wave forces
Generation	3	type of formulations
f	0.067	coefficient for bottom friction (JONSWAP)
Breaking	Battjes and Janssen (1978)	determination of depth-induced wave breaking
α	1.0	coefficient for wave energy dissipation (B&J)
γ	0.73	breaker index in the B&J model
tri	true	non-linear triad wave-wave interactions
white	true	wave whitecapping
quad	true	quadruplet wave-wave interactions
ref	true	wave refraction
freq	true	wave frequency shift

One of the objects of wave modelling is to provide correct and stable wave field in the morphological grid. Using the nested wave grid, the wave module can implement it. Fig. 4.9 shows the contours of wave heights on overall wave grid at high water level, where the flow grid is shown as a blue box. There are six plots which stand for the high wave conditions in the six schematised wave directions. From the figure, the wave boundary disturbances are kept out of the morphological grid, which ensures a stable wave field for the flow computation. The waves with different incident angles gradually deflect to be perpendicular to the shoreline, e. g. the wave crests eventually move parallel to the shoreline. When the waves propagate close to the shoreline, high waves break due to shoaling and increase of steepness, then the wave heights significantly decrease. Since the longshore bathymetry is not uniform in shallow water, the breaking happens at different distances from the shoreline in different cross sections, which leads to a various wave height distribution in the surf zone.

The varying of cross-shore wave heights can reflect wave energy propagation over the bathymetry. Fig. 4.10 shows the heights of three sets of waves coming from southwest in three representative cross sections at high and low water levels. Fig. 4.11 shows other three sets of waves coming from northwest. At the same cross section, the breaking of high weaves happens farther from the shoreline than the low waves when moving onshore. For the same incident wave, the breaking happens farther from the shoreline at low water level than at high water level. For the C. S. Middle, the high waves start breaking in front of the nourishment during low water level, about 800 m from the shoreline. During high water level, the breaking location of high waves is on the offshore slope of the outer bar, about 600 m from the shoreline.

4.3.3 Energy dissipation

The decrease in wave heights is the consequence of the loss of wave energy, i. e. energy dissipation. In this study, wave energy dissipation rate is applied to determine wave force. The dissipation rate considered with the wave model includes the energy dissipation due to depth-induced breaking, whitecapping and bottom friction. The first two are applied in the top layer, and the latter is in the bottom one.

Fig. 4.12 shows tide-averaged energy dissipation rate over the mor-

Chapter 4   Hydrodynamic Modelling

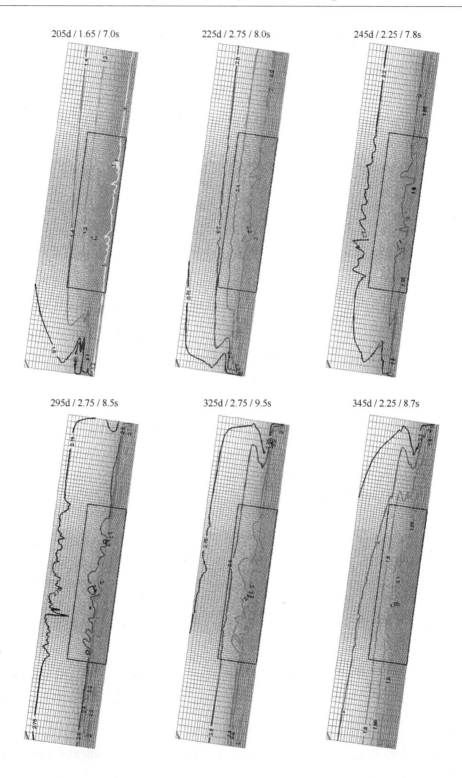

Fig. 4. 9   Wave height contours on overall wave grid at high water level

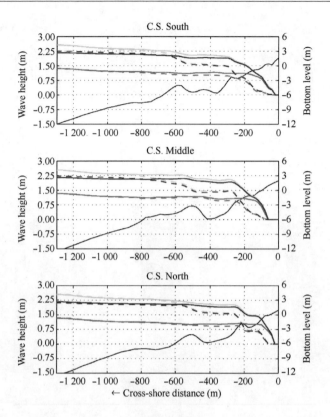

Fig. 4.10 Cross-shore wave heights for waves coming from the southwest. Solid line is at high water level, and dashed line is at low water level. Red lines are for waves with 205° incident angle and 1.65 m height; green lines for 225° and 2.75 m; blue lines for 245° and 1.25 m.

phological grid. The six plots mean the six high waves of the schematized wave conditions. The directions of the waves are pointed by a black arrow line. For the sake of comparison, all the plots use the same scale for the dissipation rate (10 ~ 100 N/m/s). The coordinate system in the plots use the location of Jan van Speijk lighthouse as the origin (0, 0), and the aspect ratio of cross-shore to longshore distances is 2. The energy dissipation rate of waves during high and low water levels are shown in Appendix Fig. A.2 and A.3.

From these figures, the deep blue parts which indicate an energy dissipation rate of less than 20N/m/s, cover the most area of the morphological grid. The deep red parts, indicating a dissipation rate of more than

Chapter 4 Hydrodynamic Modelling

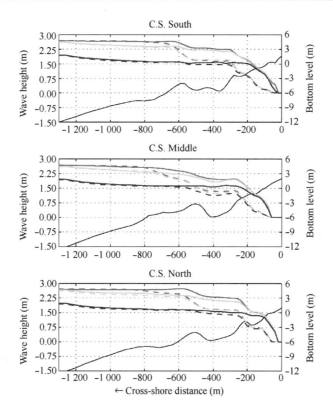

Fig. 4.11 Cross-shore wave heights for waves coming from the northwest.
Solid line is at high water level, and dot line is at low water level.
Red lines are for waves with 295° incident angle and 2.75 m height; green lines
for 325° and 2.75 m; blue lines for 345° and 2.25 m.

100N/m/s, appear on the bumps of the bathymetry. The loss of wave energy mainly happens on the longshore bars and the shoreface nourishment. In average, the inner bar contribute most to the energy dissipation, then the outer bar, and then swash bar, the last is the nourishment. During high water, the dissipation concentrates on the inner bar and the swash bar. During low water, it concentrates on the inner bar, the outer bar, and the offshore nourishment site.

4.4 Wave-current interactions

In shallow areas, the effect of waves becomes increasingly important for the current structure. Wave-current interaction plays a significant role in nearshore processes. Wave breaking is the principal driving force for currents, mean water level changes, and low frequency oscillatory motions within the surf zone, and is also believed to be of order one importance in sediment transport and large scale sand bar evolution. What follows hereafter is a brief discussion on the wave-current interaction processes implemented in Delft3D modelling system. A complete description of the wave-current interaction and its numerical implementation are given in the Delft3D-FLOW manual. The sediment transport due to wave-current interaction is described in the next section.

The following processes are presently accounted for in Delft3D-FLOW:

(1) Wave forcing due to breaking (by radiation stress gradients) is modelled as a shear stress at the water surface.

(2) The effect of the enhanced bed shear stress on the flow simulation is taken into account. The simulations presented in the model use the wave-current interaction model of Fredsøe (1984).

(3) The wave-induced mass flux is included and is adjusted for the vertically non-uniform Stokes drift.

(4) The additional turbulence production due to dissipation in the bottom wave boundary layer and due to wave whitecapping and breaking at the surface is included as extra production terms in the k-ε turbulence closure model.

(5) Streaming (a wave-induced current in the bottom boundary layer directed in the direction of the wave propagation) is modelled as an additional shear stress acting across the thickness of the bottom wave boundary layer.

In the above, the processes (3), (4), and (5) are essential if the effect of waves on the flow is to be correctly represented in 3D simulations. This is especially important for the accurate modelling of the sediment transport in a nearshore coastal zone (Walstra *et al.*, 2001; Lesser *et al.*, 2003).

The wave-current interaction are visible in wave-driven currents. Fig. 4.13 and Fig. 4.14 show the tide-averaged flow vectors for the waves coming from different directions. From the figures, the directions of longshore

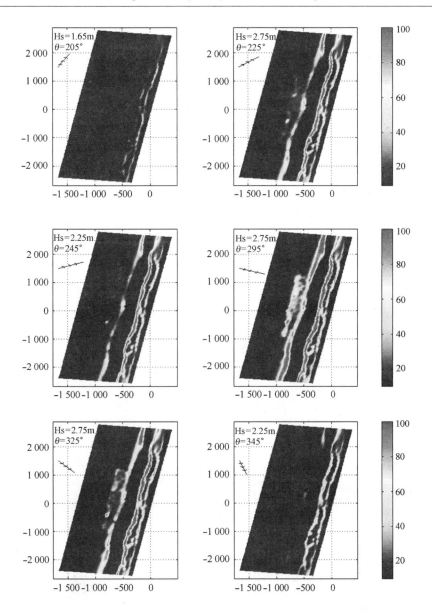

Fig. 4.12 Tide-averaged energy dissipation rate of waves (unit: N/m/s)

currents depend on the direction of the waves. Waves coming from the northwest cause southward currents, while waves coming from the southwest cause northward currents. High velocities appear on the crests of the longshore bars, though the velocities increase significantly in the whole surf zone (within 600 m from the shoreline) with respect to the tidal current. The directions of the currents are curved by the local irregular bathymetries of surf zone, and is averagely directed offshore on cross-shore profile. Strong circulations can be found when waves come more perpendicular to the shoreline. The main circle currents happen at the trough between the peaks of the outer bar and the inner bar, where there are also shallower parts of the trough.

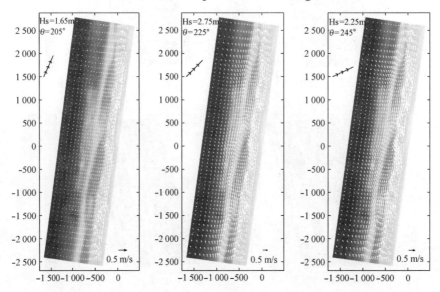

Fig. 4.13　Tide-averaged flow fields under southwest high waves

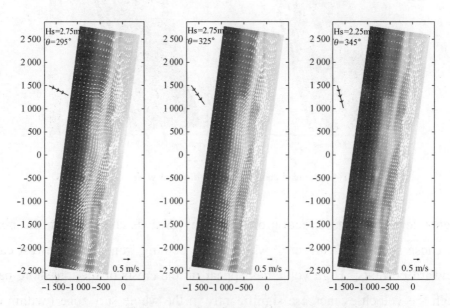

Fig. 4.14　Tide-averaged flow fields under northwest high waves

4.5 Sediment transport modelling

The sediment transport under wave-current interactions plays an important role in the morphological revolution of open coast. Egmond lies in a coastal area, where the water movement consists of currents and waves. Therefore not only the transport caused by currents, but also the transport caused by waves have to be included in the modelling of the Egmond shoreface nourishment. So to exactly simulate the sediment transports under different wave conditions becomes the key point in the study.

In the recent development, the sediment transport computation is fully integrated into the Delft3D-FLOW module, which is so-called "online sediment transport" (see Section 2.3). In the following subsection, the sediment transport formula (van Rijn, 1993) implemented in Delft3D and in the present study is briefly introduced. Then the setup of sediment transport modelling is described.

To compare with the previous study (van Duin, 2002) which used Bijker's formula (1971) and run in "offline" approach (Delft3D-TRAN module), the "online" sediment transport capacity of Bijker's is re-calculated within the Egm2004 model. For the clarity reason, the previous study is named as 2002-Bk2DH and the new case is given a name 2004-Bk2DH. Furthermore, the "online" sediment transport capacity of van Rijn's formula of Egm2004 is computed, which is named as 2004-VR2DH. At last, the results of 2DH transport modelling (online) of Bijker's and van Rijn's are discussed, with comparison to the results of Egm2002 (Bijker's formula, "offline") and the UNIBEST profile modelling (van Duin, 2002; Wiersma, 2002). Table 4.9 summaries the differences between the above mentioned computation cases.

Table 4.9 Sediment transport modelling cases

Name	Dim	Formula	Transport parameter setting
Egm2002-Bk2DH	2DH	Bijker 1971	(van Duin, 2002), offline
Egm2004-Bk2DH	2DH	Bijker 1971	same setting as Egm2002, online
Egm2004-VR2DH	2DH	van Rijn 1993	default settings, online

In this study, van Rijn (1993) is the transport formula used in morphological simulations, and Bijker (1971) is used for reference since it has been used in the Egm2002 model. If not specified, all the transport results in this section come from van Rijn (1993). The intercomparison of Bijker's and van Rijn's is beyond this study. All the factor settings specified for Bijker's are taken directly from the Egm2002 model. Table 4.10 summaries these factors, more information see (van Duin, 2002).

Table 4.10 Parameter settings for sediment transport modelling
(Egm2002-Bk2DH and Egm2004-Bk2DH)

Parameter	Value	Unit	Description
ρ_s	2 650	kg/m^3	sediment density
D_{50}	0.2	mm	medium grain size
D_{90}	0.3	mm	D_{90} grain size
BS	5	-	coefficient b for shallow water
BD	2	-	coefficient b for deep water
γ_c	0.005	m	roughness height for currents
ε	8.0	-	porosity
ω	0.023	m/s	particle fall velocity

4.5.1 Transport formula and parameter settings

At the core of the sediment transport model is an approximation method of the van Rijn (1993) formulations. The method, motivated by the need to reduce computational efforts, allows for 3D morphodynamic modelling in large spatial scale (10 to 100km) and temporal scale (years to decades). Here a short overview of the model formulations is presented with the emphasis on the formulae with user-specified factors calibrated during the sediment transport modelling and the morphodynamic simulations.

The total transport option is used in this study. The total sediment transport q (kg/m/s) is determined by the bedload transport and the suspended load transport q_s. The bedload vector due to both current and wave effects (including wave asymmetry) represents a current-related contribution (in the current direction) and a wave-related contribution ($q_{b,w}$ in the wave direction, following or opposing depending on conditions). The suspended load transport represents the current-related contribution due to advective processes ($q_{s,c}$ in the current direction) and the wave-related contribution mainly due to wave asymmetry effects ($q_{s,w}$ in the wave direction, always onshore).

In the online sediment version of Delft3D-FLOW, the transport of the suspended sediment is computed over the entire water column (from $\sigma = -1$ to $\sigma = 0$). However, for sand sediment fractions, van Rijn regards sediment transported below the reference height, a, as belonging to "bedload" sediment transport which is computed separately since it responds almost instantaneously to changing flow conditions and feels the effects of bed slope. So the wave-related suspended transport $q_{s,w}$ is included in the bedload transport vector. The three transport contribution, $q_{b,w}$ and are combined and transformed to the grid coordinate system in M and N direction:

$$q_{b,M} = f_{BED}\left[\frac{u_{b,M}}{\bar{u}_b}|q_{b,c}| + (f_{BEDW}q_{b,w} + f_{SUSW}q_{s,w})\cos\phi\right] \quad (4.5)$$

$$q_{b,N} = f_{BED}\left[\frac{u_{b,N}}{\bar{u}_b}|q_{b,c}| + (f_{BEDW}q_{b,w} + f_{SUSW}q_{s,w})\sin\phi\right] \quad (4.6)$$

where f_{BED}, f_{BEDW}, and f_{SUSW} are user-specified calibration factors, which allow users to adjust the overall significance of each transport component. $u_{b,M}$, $u_{b,N}$ and $\bar{u}_b$ (m/s) are Eulerian velocity components in the bottom computational layer, and ϕ (°) is the local angle between the direction of wave propagation and the computational grid. The magnitude and direction of the bedload transport vector can be adjusted for bed slope effects, see Delft3D-FLOW user manual for more details.

The current-related suspended load transport is defined as the transport of sediment particles by the time-averaged current velocities, which is calculated with multiplication of the velocity profile and the concentration profile. The concentration profile, c, is obtained by solving the well-known advection-diffusion equation. For more details, see van Rijn (1993; 2000).

$$q_{s,c} = \rho_s \int_{z_a}^{h} cu\,dz \qquad (4.7)$$

$$c_a = f_{sus} 0.015 \frac{D^{50}}{Z_a} \frac{T^{1.5}}{D_*^{0.3}} \qquad (4.8)$$

where, $q_{s,c}$ is current-related suspended load transport ($m^3/m/s$), ρ_s is sediment density (kg/m^3), za is reference height (m), h is water depth (m), c is time-averaged concentration (kg/m^3), u is time-averaged current velocity (m/s), z is height above the bed (m), c_a is reference concentration at height za (kg/m^3), f_{SUS} is user-specified multiplication factor, T is dimensionless bed-shear stress parameter, D_* is dimensionless parameter.

All transport contributions mentioned above are time averaged over the wave period. In order to accord with the calibrating data, the user-specified transport or morphological factors should be tuned before the implementation of morphological scenarios. This work is carried out in the next chapter. But in this section, all the user-specified factors are set to their default values, see Table 4.11.

Table 4.11 Parameter settings for sediment transport modelling (Egm2004-VR2DH)

Parameter	Value	Unit	Description
ρ_s	2 650	kg/m³	sediment density
C_D	1 600	kg/m³	dry bed density
D_{50}	0.2	mm	medium grain size

To be continued

Continued from Table 4.11

Parameter	Value	Unit	Description
D_{90}	0.3	mm	D_{90} grain size
f_{BED}	1.0	-	multiplication factor bed-load transport vector magnitude
f_{SUS}	1.0	-	multiplication factor suspended sediment reference concentration
f_{BEDW}	1.0	-	wave-related bed-load sediment transport factor
f_{SUSW}	1.0	-	wave-related suspended sediment transport factor

Sediment transport computation is executed in 2DH mode. The sediment type is sand with a medium diameter 0.2 mm. The initial conditions for the sediment fractions are handled in exactly the same manner as those for any other conservative constitute in the Delft3D system. In practical applications the non-cohesive sediment sand concentrations adapt very rapidly to equilibrium conditions, so an uniform zero concentration for the non-cohesive sediment fractions is usually adequate to satisfy computation accuracy (Lesser et al., 2003). In this study, an uniform zero concentration for the sand is used.

Boundary conditions also must be prescribed for sediment transport performance. At the water surface boundary, the vertical diffusive flux for the sand is set to zero. The exchange of material in suspension and the bed is modelled by calculating the sediment fluxes from the bottom computational layer to the bed, and vice versa. These fluxes are then applied to the bottom computational layer by means of a sediment source and/or sink term in each computational cell. The calculated fluxes are also applied to the bed in order to update the bed level. But in this section, the aim is to evaluate sediment transport capacities in the area of interest, so the bed level update is blocked. Some details is given in the following chapter, for more information see the Add-ons of Delft3D-FLOW user manual.

At all inflow boundaries, the boundary conditions are required. Equilibrium sediment concentration profiles are usually specified at the open inflow boundaries to avoid high accretion or erosion rates near the model boundaries. No boundary conditions are prescribed at outflow boundaries.

Some parameters related to sediment transport are summarised in Table 4.11, where the user-specified transport factors are given in default values. In the next chapter, these factors will be calibrated by profile models and then be deployed in the morphological simulations.

4.5.2 Sediment transport in 2DH mode

Prior to discussing the transport due to wave-current interactions, the transport due to tide is first briefly described. As mentioned in the preceding sections, the tide residual flow is slightly northward, thus the tide-averaged transport should be also northward. Fig. 4.15 shows the tide-averaged transports at three representative cross sections. The tide induced transport are dominant outside the surf zone (deeper than 8 m water depth), and the transport in the surf zone can almost be neglected relatively. In average, the longshore transport is in the order of 10^{-4} m^3/s. The cross-shore transport with

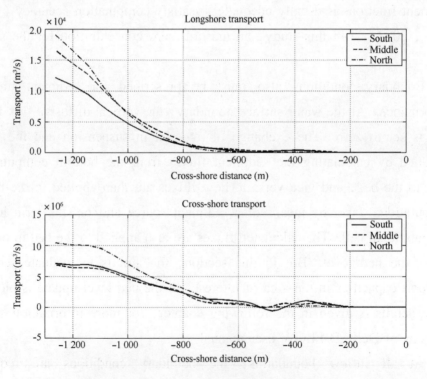

Fig. 4.15 Tide-averaged sediment transport on the cross section South, Middle and North (Egm2004-VR2DH). No waves included. Positive means northward longshore transport in the upper plot and onshore transport in the lower plot.

an onshore direction is in the order of 10^{-6} m^3/s. The magnitude of longshore transport on the middle cross section is less than that on the north section, and the south section has the smallest magnitude. It can be explained that the nourishment decreases the upper-stream velocities. A gradient of longshore transport forms between the south and the north section, which could cause erosion on the nourishment. As for the cross-shore transport, the middle section is lower than the other two sections. Since the nourishment also decrease the onshore velocities, the transport is partly blocked. The trend of cross-shore transport outside the surf zone will cause sediment moving onshore.

The sediment transports due to wave-current interactions are quite different from the tide induced transport. Fig. 4. 16 presents the transports due to all input wave conditions, twelve sets of waves with different directions and heights. The upper two plots describe the transport due to high waves, and the lower two ones are due to average (low) waves. The magnitudes of these transports are averaged over the tide cycle and over the space of the morphological grids. The peaks of longshore and cross-shore transports on the cross section locate on the trough between the outer bar and inner bar, close to the inner bar. The magnitudes of the transports are dependent on the wave heights. The higher waves cause higher transports. The cross-shore transports are always onshore under the default settings of the transport factors (see Table 4. 11). The inner bar, the outer bar and the nourishment get higher onshore transport, which will result in erosion on the outer slopes of them and accretion on the inner slopes.

To compare with the results of previous study (Wiersma, 2002; van Duin, 2002), the tide-averaged and cross-section (C. S. South in Fig. 4. 2) integrated longshore transport (m^3/s) for each wave group is listed in Table 4. 12. Table 4. 13 describes the transported volume (m^3) accu-mulated over morphological duration for each wave group. Fig. 4. 17 gives the expression of all the results listed in the above tables. The results of UNIBEST and Egm2002-Bk2DH are taken directly from the previous study (Wiersma, 2002)

. From Fig. 4.17, the higher transport capacities are caused by the higher waves, especially the waves coming from 225°N (southwest) and 325°N (northwest). Compared among four modelling cases, UNIBEST almost has the largest value for each wave condition than other modelling cases, except for 345h in which Egm2004-VR2DH is the largest. Egm2004-Bk2DH has less transports than Egm2002-Bk2DH, except for the wave condition 325h. For the wave condition 325h, the transports of Egm2004-Bk2DH and Egm2002-Bk2DH are equal to each other approximately. Egm2004-VR2DH has different performances for northward and southward transport. For the northward transport, it has the least values between four cases in most wave conditions; however, it is always larger than Egm2002-Bk2DH and Egm2004-Bk2DH for southward transport. On the whole, the northward transport is more consist than for the southward transport.

Table 4.12 Tide-averaged and cross-section† integrated longshore transport per wave condition

Southwest waves	205a	205h	225a	225h	245a	245h
UNIBEST[1]	3.15	39.9	21.4	191	13.1	79.2
Egm2002-Bk2DH[2]	1.45	19.5	10.3	120	8.98	58.8
Egm2004-Bk2DH[3]	0.51	10.76	6.36	101.19	6.63	45.05
Egm2004-Bk2DH[4]	1.86	12.24	6.75	83.41	5.22	33.61
UNIBEST	-13.4	-141	-23.2	-265	-2.29	-49.0
Egm2002-Bk2DH	-4.36	-64.4	-7.83	-91.4	0.04	-31.8
Egm2002-Bk2DH	-0.96	-47.07	-1.98	-91.36	0.23	-21.13
Egm2002-Bk2DH	-7.42	-71.82	-14.58	-146.41	-1.35	-56.42

† indicates C. S. South shown in Fig. 4.2, transport unit: 10^{-3} m³/s;

‡ 205 ~ 345 are wave incident angles with respect to the north;

‡ "a" means averaged wave; "h" means high wave;

[1] takes the results directly from previous study (van Duin, 2002);

[2] takes the results directly from previous study (Bijker formula and "offline");

[3] uses Bijker formula and "online" approach;

[4] uses van Rijn formula and "online" approach.

Chapter 4 Hydrodynamic Modelling

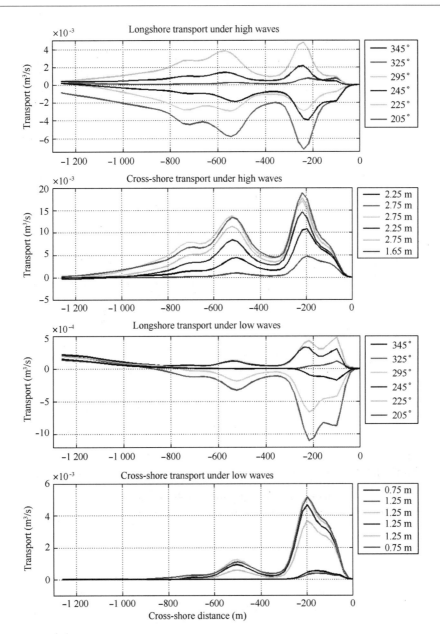

Fig. 4.16 Tide-averaged sediment transport on cross-section under wave-current interactions (Egm2004-VR2DH). The magnitude is averaged over the tide cycle (time) and over the longshore distance (space).

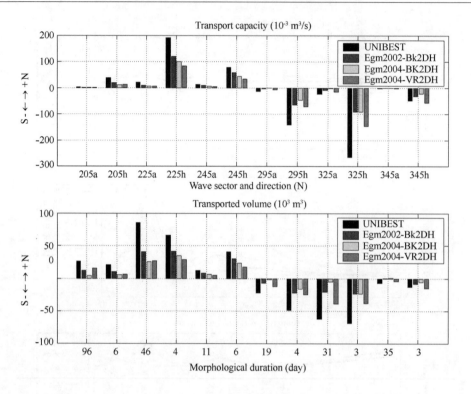

Fig. 4.17 Tide-averaged and cross-section integrated longshore transport capacity and transported volume accumulated over morphological duration per wave group

Table 4.13 Total longshore transported volume per wave condition accumulated over morphology duration

Southwest waves	205a	205h	225a	225h	245a	245h
UNIBEST	26.1	20.7	85.1	66.0	12.5	41.1
Egm2002-Bk2DH	12.0	10.1	40.9	41.5	8.5	30.5
Egm2004-Bk2DH	4.2	5.6	25.3	35.0	6.3	23.4
Egm2004-Bk2DH	15.4	6.3	26.8	28.8	5.0	17.4
Northwest waves	**295a**	**295h**	**325a**	**325h**	**345a**	**345h**
UNIBEST	-22.0	-48.7	-62.1	-68.7	-6.9	-12.7
Egm2002-Bk2DH	-7.2	-22.3	-21.0	-23.7	0.1	-8.2
Egm2002-Bk2DH	-1.6	-16.3	-5.3	-23.7	0.7	-5.5
Egm2002-Bk2DH	-12.2	-24.8	-39.1	-37.9	-4.1	-14.6

Although the transport capacities of the low waves are much less than the high waves, but the transported volumes over the morphological durations of the low waves can not be neglected, see the lower plot of Fig. 4.17. In UNIBEST, the average wave of 225°N (225a) contributes more to the total net transport than other waves. In Egm2002-Bk2DH, main contributions come from the wave conditions of 225a and 225h and they almost contribute the same volumes to the total transport. In Egm2004-Bk2DH, 225h has the maximum transported volume. For Egm2004-VR2DH, the peak values do not appear in northward transported volumes but southward ones. The wave 325a has the maximum, and 325h is slightly less.

The net transport per wave direction (summation of average wave and high wave with the same incident direction) is described in Table 4.14. And the total net transport over the morphological duration is listed in the last column of the table. Fig. 4.18 displays these sums. The results of UNIBEST show that the main transports is caused by the waves directed 225° and 325° and the former (northward transport) is larger than the latter (southward transport). For Egm2002-Bk2DH and Egm2004-Bk2DH, the main transports come from 225°. Similar to UNIBEST, the main transports of Egm2004-VR2DH also come from 225° and 325°, but other than UNIBEST, the northward transport is less than the southward transport. Totally, the net transports of UNIBEST, Egm2002-Bk2DH, and Egm2004-Bk2DH are northward, but the magnitude of UNIBEST is half of Egm2002-Bk2DH, which are 30.2×10^3 and 61.4×10^3 m^3 respectively. The result of Egm2004-VR2DH is southward directed, though its magnitude is more or less equal to the result of UNIBEST. It should be kept in mind that the transport factors of Egm2004-VR2DH are all set to their defaults at present. The differences between these modelled results are discussed in the coming subsection.

Shore Nourishment and Morphodynamic Modelling

Table 4.14 Net transport over morphological duration per wave direction (Unit: 10^3 m^3)

Wave direction (°N)	205	225	245	295	325	345	Total
UNIBEST	46.8	151.1	53.5	-70.7	-130.8	-19.6	30.2
Egm2002-Bk2DH	22.1	82.4	39.0	-29.4	-44.7	-8.1	61.4
Egm2004-Bk2DH	9.8	60.2	29.7	-17.8	-29.0	-4.8	48.1
Egm2004-VR2DH	21.8	55.7	22.4	-37.0	-77.0	-18.7	-32.9

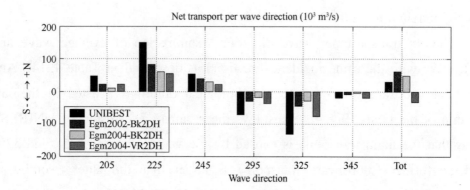

Fig. 4.18 Net transport over morphological duration per wave direction and total net transport of all wave directions

4.5.3 Discussions on net transport

The longshore transport at Egmond (Section 38km of the Dutch coast) is dominated by the offshore wave conditions and sensitive to the wave boundary conditions. Although there are no man-made structures in the area of interest, the site is located northward 17km of IJmuiden (Section 55km) and southward of Seawall (Section 21km). In the study of sand budget and coastline changes of the central Dutch coast (van Rijn, 1995), it is shown that North of IJmuiden the net transport rate is directed southward in Section 35 ~ 55km and northward again in Section 0 ~ 35km. The net southward longshore transport north of IJmuiden is related on the presence of the harbour dam reducing the wave energy coming from southwest directions. The influence of the dam

Chapter 4 Hydrodynamic Modelling

appears to extend over a rather long distance (about 20 km or 8 times the dam length). From Fig. 4.19 which is cited from the above study, the yearly-averaged net transport in the surf zone of Egmond (38km) is close to zero. It is possible for the transport in northward direction or in southward direction, according to the results of different authors.

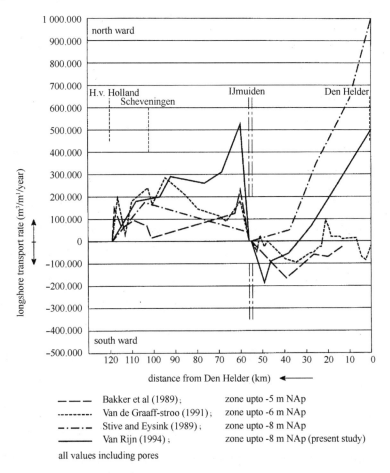

Fig. 4.19 Yearly-averaged net longshore transport in surf zone along the Dutch coast (van Rijn, 1995). Egmond is located on the distance 38 km from Den Helder.

As mentioned above, the distinguish abilities between Egm2002-Bk2DH and Egm2004-Bk2DH are that the computed areas and boundary conditions of both grids are different, and the sediment transport computations use different approach ("offline" versus "online"). From Fig. 4.18 the magnitudes of the transport for each wave direction of Egm2004-Bk2DH are 44% ~ 76% of

89

those of Egm2002-Bk2DH. The total net transport of Egm2004-Bk2DH is equal to 79% of Egm2002-Bk2DH. After comparing the tide-averaged and cross-shore integrated longshore transport in Table 4.12, the least ratios between Egm2004-Bk2DH and Egm2002-Bk2DH are 22% (295a) and 25% (325a). Relatively, the results of high waves in both cases are closer than low waves. The results of Egm2002-Bk2DH are taken directly from the previous study. The comparison between these two cases is reasonable, due to the differences between them.

Egm2004-Bk2DH and Egm2004-VR2DH almost share the same input conditions, except for the transport formulas. Furthermore, both transport formulas have their specified factors, which cause it difficult to compare the results of their. The intercomparison of these transport formulas is beyond the scope of this study, more detailed information on the intercomparison of different sediment transport models can be referred in (Davies *et al.*, 2002).

Noted that all the transport factors in Egm2004-VR2DH at present are set to their default values, see Table 4.11. Sediment transport processes depend closely on local hydrodynamic and morphohdynamic conditions. The corresponding transport factors should be calibrated through sensitivity analysis. The same transport formula are used in this study and UNIBEST, in spite of the fact that there are slight difference between the implementations in two models (van Rijn, 2000; Walstra, 2001). The net transport in this study will be calibrated against the results of UNIBEST, since the schematised wave conditions are based on the predicted net transport of UNIBEST. This work will be done in the next chapter.

4.6 Conclusions

The present Egmond 3D model consists of two grids used respectively for the FLOW and WAVE modules. The flow (morphological) grid covers the area of $1\ 300 \times 5\ 200\ m^2$ which is nested within the wave grid of $2\ 400 \times 14\ 900\ m^2$ The vertical profile of the flow grid is separated into 11 layers for

3D simulation. The bottoms of the grids use the measured bathymetry dated on 1 September 1999.

The schematised tide adopts the result of previous study. However the boundary conditions have to be regenerated to suit the new grid. Riemann variants are derived as the lateral boundary conditions of the model. The calibration against the previous study shows that the new model not only well reproduces the tidal currents and makes more stable results.

The wave computation of the model is based on the default settings of the SWAN system. The boundary conditions (the schematised wave conditions) are also copied directly from the previous study. The model can reflect the offshore wave propagation over the area of interest. The output of the wave computation provides an orderly wave field for the flow grid. Most of wave breaking happens on the longshore bars and the nourishment. The wave energy dissipation mainly concentrates on the longshore bars. Wave-current interactions significantly change the flow pattern within the surf zone.

The formula of sediment transport used in the model is van Rijn's, which includes the transport caused by both currents and waves. The local sediment transport is sensitive to the wave boundary conditions. Moreover, the computed transport relies on the settings of the transport factors used in the formula. In this chapter, all the transport factors are set to their defaults. There are larger discrepancies between the present study and the previous studies on the net transport. The reasons may be caused by the different formulae adopted in the models and the settings of the transport factors.

Chapter 5

Morphodynamic Validation

This chapter focuses on the morphodynamic modelling. The implementation of "online sediment transport" approach has been introduced briefly in Chapter 2 (Section 2.3). Following this approach, the morphological scenario is first discussed in Section 5.2, especially on the determination of the morphological acceleration factors.

The net transport of the modelled area is sensitive to the settings of transport factors (based on the van Rijn 1993 formula). These factors should be adjusted according to the schematised net transport, i.e. the results of UNIBEST. Due to the considerable time efforts in running fully 2DH and 3D area models, Delft3D models running in profile mode are applied to calibrate the transport factors in Section 5.3. Corresponding to 2DH and 3D area models, the profile models are in 1DH and 2DV modes respectively. In these profile models, the wave computation is performed by the Surf Zone Wave (SZW) model, i.e. roller model with Snell's, which is fully integrated into the Delft3D-FLOW module. Except for modelling on sediment transport, the profile models are also used to evaluate morphological evolutions. With the help of statistic method (Brier Skill Score), the final calibrated transport factors are chosen based on the modelled results.

To compare with the previous morphological hindcast modelling Egm2002-Bk2DH (van Duin, 2002), two 2DH morphological simulations with "online transport approach" are carried out following the transport tests mentioned in the preceding chapter. The calibrated morphological factors by the 2DV profile model are eventually used to execute three-dimensional area morphodynamic modelling in the area of interest. Two cases with different transport factors are executed in Section 5.4. The modelled results are compared with the measured data and other simulation results mentioned above. These results will also be discussed in different aspects.

5.1 Morphological scenario

The morphological evolution of the Egmond coastal area concerns the fully coupled activities of waves, flow, and sediment transport and bed level variation. In Delft3D this fully coupled dynamic system is decomposed in separate systems (i.e. the WAVE module and the FLOW module) which are operated sequentially but are using each other's results (see Fig. 2.4, the diagram of Delft3D morphodynamic procedure). The Delft3D-FLOW is the steering module to control the execution of the morphological simulations. It controls the order in which WAVE and FLOW modules are activated and how data communication is organized. Furthermore, it has a number of options to stop a process, more details see Delft3D-FLOW and Delft3D-WAVE user manuals. The simulation procedure is organized by FLOW module according to user-defined process order of flow and wave.

5.1.1 Morphodynamic simulation procedure

Sediment transport is fully included in the FLOW module with the online sediment transport approach which can update the bed level changes each time step and feed back to hydrodynamic simulations simultaneously. So the process tree is simply composed by calling the FLOW and WAVE modules in turn, see Fig. 2.4. However, large bed level changes in the FLOW module may cause computation unstable. In order to keep the hydrodynamic and morphodynamic simulations ongoing stably, it is important to determine the morphological acceleration factors companied with effective simulation time, which is discussed in the following subsection.

In the morphological simulations of this study, each time twelve wave conditions with six different incident directions are calculated one by one. Delft3D-FLOW distributes the simulation time for the flow computations coupled with different wave conditions and morphological acceleration factors. Fig. 5.1 describes the simulation procedure. Delft3D-FLOW runs in restart mode, i.e. it initialises the simulation based on the results of the last Delft3D-FLOW run. For each new start, it first reads the restart file (communication file) and new wave condition. Each run produces a new restart file, besides the flow results (include sediment transport and morphology changes) and the wave results. Such procedure has the following advantages: ① if something goes wrong for one wave condition, the simulation does not have to be repeated for the whole calculation; ② more results can be stored; ③ the procedure does not become more complex as more wave conditions are used.

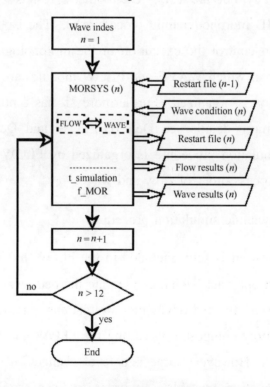

Fig. 5.1 Morphological simulation procedure

A question appears that to what extent the order of the input wave conditions will affect the final results. The wave chronology may have an impact on the morphological evolution, and some studies have been carried out to focus on this problem (Southgate, 1995). An ascending and a descending wave height orders has been used to predict the morphological changes of different time span in a process-based cross-shore modelling on the Egmond shore-face nourishment (Grasmeijer, 2002). He found the differences due to the effect of changing the wave chronology were small, whether in the short-term or in the medium-term. Van Duin (2002) made a sensitivity run using a random order of input wave conditions and compared with a hindcast results, which displayed that the order of wave conditions is of no inpertance in 2DH morphological modelling on the Egmond shoreface nourishment. A Delft3D modelling on the Terschelling shoreface nourishment (Grunnet *et al.*, 2004) also showed that changing the order of the forcing conditions revealed negligible changes in cumulative predicted bed level changes at the end of the simulation period.

These findings suggest that nearshore profile behaviour depends on the cumulative amount of energy input rather than the sequence of events. So in this study, the wave chronology keeps unaltered. The wave directions change from the southwest to the northwest, and the wave heights alter from the average wave height to the high wave height (storm) in turn, i.e. the order in Table 4.7 and Table 5.1

5.1.2 Morphological acceleration factor

As mentioned above, the bed level changes are multiplied with a morphological acceleration factor, f_{MOR}. By multiplying morphological changes with this factor the computation time can be significantly reduced; however numerical instability may result in when f_{MOR} exceeds a certain value which causes higher bed level (water depth) gradients and therefore effects hydrodynamic computation. It has been demonstrated that for simple cases very high morphological factors can be used without significantly changing the solution. For tidal and wave-driven situations results have been presented indicating that values in the order of 50 ~ 100 are viable in the study of Lesser *et al.* (2003). In this reference, it was also stated that the use of even a rather high morphological acceleration factor has little impact on the development of the morphology in some situations (e.g. coastal area). But the authors stressed that appropriate morphological acceleration factors must be chosen and tested on a case-by-case basis.

In the present study, we simply use the ratio of the simulation time (one or more tide cycles) and the real morphological time (days or months) to determine the morphological acceleration factor. Equation 2.6 is then modified to

$$f_{\text{MOR}} = \frac{T_{\text{morphology}}}{T_{\text{simulation}}} \quad (5.1)$$

$$T_{\text{simulation}} = nT_{\text{tide}}, \quad (n = 1,2,\ldots) \quad (5.2)$$

where,

f_{MOR}——morphological acceleration factor;

$T_{\text{morphology}}$——schematised wave duration acting on morphological evolution;

$T_{\text{simulation}}$——computational time scaled to morphological evolution;

N——number of tide cycles used to simulate morphological evolution;

T_{tide}——tide cycle period, equals 12.5 hours in present study.

Sensitivity runs show that the value of the factor exceeds 50 will cause the computation unstable in the present application. So f_{MOR} remains below 50 in the morphodynamic simulations of this study. For each morphology time, a factor as large as possible is calculated by using the number of tide cycle n as small as possible. The factors used in the morphological simulation scenario are summarized in Table 5.1. In the table, MORSTT is the computational spin-up time.

Table 5.1 Simulation time and morphological acceleration factors

Wave order	Direction (°N)	H_s (m)	$T_{\text{morphology}}$ (days)	$T_{\text{simulation}}$ (hours)	Tide n	MORSTT (hours)	f_{MOR}
1	205	0.75	96	50.0	4	3.0	46.08
2	205	1.65	6	12.5	1	2.5	11.52

To be continued

Continued from Table 5.1

Wave order	Direction (°N)	H_s (m)	$T_{morphology}$ (days)	$T_{simulation}$ (hours)	Tide n	MORSTT (hours)	f_{MOR}
3	225	1.25	46	37.5	3	2.5	29.44
4	225	2.75	4	12.5	1	2.5	7.68
5	245	1.25	11	12.5	1	2.5	21.12
6	245	2.25	6	12.5	1	2.5	11.52
7	295	1.25	19	12.5	1	2.5	36.48
8	295	2.75	4	12.5	1	2.5	7.68
9	325	1.25	31	25.0	2	3.0	29.76
10	325	2.75	3	12.5	1	2.5	5.76
11	345	0.75	35	25.0	2	3.0	33.60
12	345	2.25	3	12.5	1	2.5	5.76
Total			264	237.5	19	31.5	

In practical computation, a hydrodynamic simulation will take some time (i.e. so-called spin-up time) to stabilise after transitioning from the initial conditions to the (dynamic) boundary conditions. It is likely that during this stabilisation period the patterns of erosion and accretion that take place do not accurately reflect the true morphological development and should be ignored. The time interval for this period, MORSTT, is used in Delft3D-FLOW. During the MORSTT interval all other calculations will proceed as normal except that the effect of the sediment fluxes on the available bottom sediments will not be taken into account. After this specified time interval, the morphological bottom updating will begin. In the present study, The Delft3D-FLOW simulation becomes stable in 150 minutes when it runs in restart mode. Therefore, the MORSTT is set to 2.5 hours at least. To be in harmony with the updating of wave computation called by Delft3D-FLOW, some MORSTTs are set to 3 hours. The values of MORSTT are listed in Table 5.1 too. Then the flow simulation time coupled with each wave condition has to be added by

MORSTT, so it is not exactly equal to the value calculated by Equation 5.1 but the sum of $T_{simulation}$ and MORSTT.

In the morphodynamic simulation, the wave computation is updated each hour, so the WAVE module will be called 269 times (237.5 + 31.5) following the morphological simulation procedure, see Table 5.1. According to the test runs (Intel P4 2.4GHz, 256M MEM), it takes about 3 to 3.5 hours to finish one tide cycle simulation in 3D mode. So totally it takes at least 57 hours to fulfil a whole morphological simulation. If the interval for wave updating is set to 0.5 hours, the total computation time is approximately added 10 more hours (0.5 hour per tide cycle). Test runs also showed that there were no significant differences between 1 hour and 0.5 hour wave updating intervals. However it means considerable time efforts to run more fully morphological simulations. For the sake of simplicity and practice, Delft3D models running in profile mode are set up in the next section to calibrate the related morphological factors.

5.2 Calibration on transport factors

Profile models have been used for the estimation of longshore transport rates, the development of cross-shore bed profiles due to the sand nourishments and the prediction of dune erosion volumes (van Rijn *et al.*, 2001, 2003; Grasmeijer and Walstra, 2003). The profile models applied herein is based on Delft3D, which is actually a cross section extracted from the corresponding 2DH and 3D area model. Wave simulation in the profile models uses Surf Zone Wave model (roller model with Snell's, see Section 2.3). Except for evaluating longshore sediment transport (based on van Rijn 1993 formula), the profile models are also used to evaluate bottom changes in this section.

The calibrated transport factors will be eventually applied in morphodynamic simulation of area models (2DH or 3D). The profile models discussed in this section can be considered as the simplified versions of the corresponding 2DH and 3D area models. The final area morphodynamic modelling is based on an assumption: the sediment transport and the morphological evolution of profile model represent the processes of corresponding area model.

5.2.1 Setup of 1DH and 2DV profile models

Similar to Delft3D area model setup, a grid should be made first for profile model in which only one longshore grid cell presents. In the present study, the middle cross section (Section Jan van Speijk, see Fig. 4.1 and Fig. 4.2) of the Egm2004 flow grid is chosen as the profile of interest. The cross shore grid cells are exactly the same as the flow grid. 1DH profile corresponds to 2DH area model, and 2DV does to 3D area model. There are no vertical grid cells in 1DH model, i.e. only 1 layer in vertical resolution, but the vertical grid resolutions in 2DV model present in multi-layer and are also based on σ-coordinate (11 layers), see Fig. 5.2.

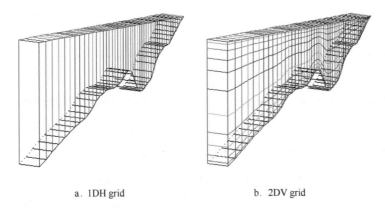

a. 1DH grid   b. 2DV grid

Fig. 5.2 Grids of the 1DH and 2DV profile models (Jan van Speijk cross-section)

The profile models are also driven by tide and almost share the same boundary conditions as their corresponding area models described in the preceding chapter. However, the tide components should be re-generated due to the changes of boundary locations. The method of generation has been described in the preceding chapter. Only one longshore grid cell presents in the profile model (only one boundary point at sea side), So there is no water level difference between the two lateral boundaries, i. e. the water level at the sea boundary has no gradient over longshore distance. The tide boundary conditions at the sea side are summarised in Table 5.2. Therefore the two lateral ends have identical Riemann conditions (water level gradient against longshore distance). The water level on the sea boundary will be simultaneously going up and down following the tide cycle.

Table 5.2 Tide components of 1DH and 2DV profile models

Angular velocity (°/hour)	Amplitude (cm)	Phase (°)
0.0	13.025	0.000
28.8	70.466	151.270
57.6	25.716	-139.970
86.4	5.061	-28.704
115.2	7.095	-6.046
144.0	0.930	134.970
172.8	1.758	133.480

Another important parameter setting is the threshold water depth for drying and flooding of the grid cell. Since there is only one longshore grid cell, the water movement is limited in a relevant small space. If the bottom updating is switched on, special attentions should be paid to the threshold depth setting. A larger or smaller value may cause the instability of the computation. Sometimes it is necessary to try a couple of values in order to get a stable running. The Delft3D-FLOW input file (rID.mdf) used by the 2DV profile model is attached in (Sun, 2004).

The boundary conditions for sediment transport are almost same as the area model. In 2DH area model, the sediment concentrations at the open inflow boundary is completely identical on the vertical section (1 layer). In 3D area model (11 vertical layers), Delft3D-FLOW allows users to prescribe the concentration at every σ-layer using a time series. A local equilibrium sediment concentration profile for multi-layer usually satisfy accuracy for sand sediment fraction (Roelvink and Reniers, 2012). So in both profile models, an equilibrium concentration profile is applied at the inflow open boundary.

5.2.2 Sensitivity runs on net longshore transport

The sediment transport computations are also steered by Delft3D-FLOW. The steering module read wave boundary conditions from the file wavecon.rID, and control the time distribution of flow simulation. Only when the sediment transport is taken into account, the bottom remains unchangeable. The transport for each wave condition is computed respectively. As a sensitivity study, six sets of transport factor combination are used to test the sediment transport. Table 5.3 summaries these combinations of factors, in which "p" means profile model and all default values are not framed. The factors were described in Table 4.11.

Table 5.3 User-specified sediment transport factors

Run name	f_{SUS}	f_{BED}	f_{SUSW}	f_{BEDW}
psw1p0	1.0	1.0	1.0	1.0
psw0p5	1.0	1.0	0.5	1.0
psw0p0	1.0	1.0	0.0	1.0
pbw0p5	1.0	1.0	0.0	0.5
ps1p5	1.5	1.0	0.0	1.0
pb0p5	1.0	0.5	0.0	1.0

5.2.2.1 Results of 1DH profile model

The computed net longshore transport of 1DH profile model is summarised in Table 5.4 and plotted in Fig. 5.3. The tide-averaged transport on the cross-shore section for each sensitive run is attached in Appendix Fig. A.4 ~ Fig. A.9. The tide-averaged and cross-section integrated sediment transport, the transported volume accumulated over morphological duration for each wave condition, are also shown in Appendix Fig. A. 10 and Fig. A. 11. For comparison, the results of UNIBEST are plotted in each figure mentioned here.

Table 5.4 Net longshore transport per wave sector of 1DH profile model [Unit: 10^3 m^3]

Run names	Wave Direction (°N)						Total
	205	225	245	295	325	345	
UNIBEST	46.8	151.1	53.5	-70.7	-130.8	-19.6	30.2
psw1p0	18.5	96.3	43.0	-48.7	-115.1	-25.3	-31.2
psw0p5	14.3	81.1	38.0	-37.5	-85.7	-17.0	-6.8
psw0p0	10.1	66.0	33.0	-26.3	-56.4	-8.8	17.6
pbw0p5	9.4	62.3	31.7	-23.7	-50.5	-7.2	22.0
ps1p5	14.3	94.9	48.1	-36.9	-78.5	-11.6	30.2
pb0p5	9.9	65.6	32.9	-26.4	-56.2	-8.8	17.0

Based on the tide-averaged sediment transport, the net longshore transport for different combinations of transport factors is obtained by multiplying transport capacities and morphological durations. The upper plot of Fig. 5.3 shows the net transport for different f_{SUSW} values (0.0, 0.5, 1.0), and other factors using default values (see Table 5.3). From the figure, the direction of net transport is sensitive to the f_{SUSW} setting. Larger f_{SUSW} value brings more southward transport. When $f_{SUSW} = 0$, the direction of net transport changes to the northward, which is consistent with the result of UNIBEST. The lower plot of Fig. 5.3 shows the net transport for the sensitivity runs pbw0p5, ps1p5, pb0p5. These cases have an identical f_{SUSW} value of 0. According to the figure, all the cases have northward transport, in which ps1p5 has the maximum magnitude and is closer to the result of UNIBEST. The values of psw0p0 and pb0p5 are nearly the same as each other, which indicates that the f_{BED} is insensitive in this situation.

In relation to the default values (psw1p0), a larger f_{SUS} corresponds to

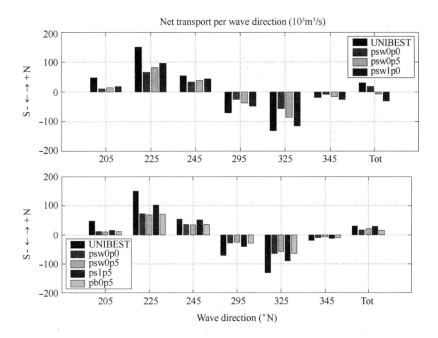

Fig. 5.3 Net longshore transport per wave sector of 1DH profile sensitivity runs. The results of UNIBEST are taken directly from the previous study. In the upper plot f_{SUSW} changes from 0.0 to 0.5, 1.0, and other factors are set to the default 1.0. In the lower plot, $f_{SUSW} = 0$ and other factors is 1.0 if not specified. For comparison, UNIBEST and psw0p0 in the lower plot repeat the same results as the upper plot.

more offshore current-related suspended load and a smaller f_{SUSW} corresponds to less onshore wave-related suspended sediment transport, a smaller value of f_{BED} corresponds to less onshore bed-load transport. These correspondences can be proved in the tide-averaged transport on cross-section, see Appendix Fig. A.4 ~ Fig. A.9. Comparing the case of psw0p0 (Fig. A.6) with psw1p0 (Fig. A.4), the apparent difference is offshore transport happens in the case psw0p0. The offshore transport takes place at the outer bar and the inner bar of the profile. It is larger with higher waves, but is also dependent on wave incident angles. For all the cases with the f_{SUSW} value of 0, offshore transport can be observed on the profile. However, the wave-related suspended transports are dominant if $f_{SUSW} = 0.5$ or $f_{SUSW} = 1.0$, which hardly have offshore transport capacity.

The net transport, the longshore transport and cross-shore transport on the profile are correlate. The change of any one factor may cause the global changes of the longshore transport and cross-shore transport and therefore the net transport. If no offshore transport is present, the sand would be always pushed onshore and the nearshore transport mechanism could not be modelled properly. Such point has been proved in test runs, so the morphodynamic simulations which are discussed hereafter do not take the cases psw1p0 and psw0p5 into account any more.

5.2.2.2 Results of 2DV profile model

Table 5.5 lists the computed net longshore transport of 2DV profile model, and these results are also presented in Fig. 5.4. The tide-averaged transport on the cross section for each sensitivity run is attached in Appendix Fig. A.12 ~ Fig. A.17. In addition, Fig. A.18 expresses the tide-averaged and cross-section integrated sediment transport, and Fig. A.19 shows the transported volume accumulated over morphological duration for each wave condition. The results of UNIBEST are also plotted in each figure for comparison.

Table 5.5 Net longshore transport per wave sector of profile model
(Unit: 10^3 m^3)

Run names	Wave Direction (°N)						Total
	205	225	245	295	325	345	
UNIBEST	46.8	151.1	53.5	-70.7	-130.8	-19.6	30.2
psw1p0	19.5	102.2	45.4	-51.4	-123.2	-26.2	-33.7
psw0p5	15.3	87.1	40.4	-39.9	-93.6	-17.9	-8.6
psw0p0	11.1	71.8	35.3	-28.4	-64.0	-9.6	16.2
pbw0p5	10.3	68.1	34.0	-25.7	-58.0	-7.9	20.7
ps1p5	15.6	103.5	51.5	-40.0	-89.8	-12.7	28.1
pb0p5	10.9	71.3	35.2	-28.5	-63.8	-9.6	15.5

The results of the 2DV profile model have almost the same features as those of the 1DH profile model, except for slight differences in magnitude. In the following paragraphs, the results of both profile models are compared and analysed.

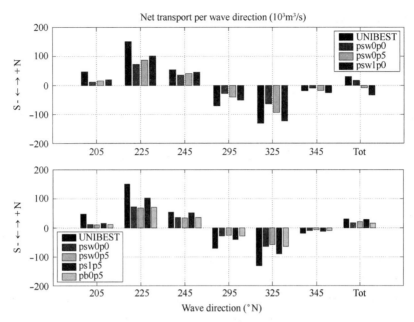

Fig. 5.4 Net longshore transport per wave sector of 2DV profile sensitivity runs. In the upper plot f_{SUSW} changes from 0.0 to 0.5, 1.0, and other factors is 1.0. In the lower plot, $f_{SUSW} = 0$ and other factors is 1.0 if not specified. The results of UNIBEST are taken directly from the previous study. For reference, UNIBEST and psw0p0 in the lower plot repeat the same results as the upper plot.

5.2.2.3 Intercomparison on sediment transport of both profile models

Firstly, we check the net longshore transport. The results of each cases in both models are quite close to each other. The magnitudes of psw1p0 and psw0p5 in 1DH model are less than those in 2DV model, while the magnitudes of other fours cases in 1DH are slightly (1.06 ~ 1.1 time) larger than those in 2DV. As mentioned before, the former two cases have southward transport and have almost no offshore transport. These phenomena may be explained by hydrodynamic conditions due to the different computational grids of the profile models.

Then, the tide-averaged and cross-shore integrated transports of both profile models are compared. Table 5.6 summaries the values. The relative difference between the transports of both models are also calculated in the table. To show the differences clearly, such differences are presented in Fig. 5.5. Some relative differences in 345a are considered as zero, i.e. there are no difference, since the absolutes of transport in 345a are very small but they can cause the relative differences become quite large.

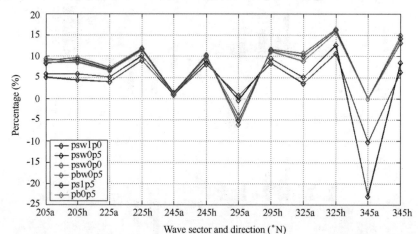

Fig. 5.5 Relative difference of tide-averaged and cross-section integrated longshore transport of 1DH and 2DV profile models. Positive means the value of 2DV is larger than 1DH, otherwise negative means 2DV is less than 1DH.

Table 5.6 Comparison of tide-averaged and cross-section integrated transport of each wave condition in 1DH and 2DV profile models

Wave(°N)		205a	205h	225a	225h	245a	245h	295a	295h	325a	325h	345a	345h
psw1p0	1DH	1.32	14.55	10.40	158.95	8.62	67.18	−8.53	−100.30	−20.50	−232.08	−1.61	−78.77
	2DV	1.39	15.22	10.83	174.72	8.70	73.11	−8.59	−109.40	−21.26	−259.64	−1.46	−84.05
	D(%)	5	4	4	9	1	8	1	8	4	11	−10	6
psw0p5	1DH	1.05	10.85	8.22	140.205	7.24	60.09	−5.56	−82.10	−13.76	−188.57	−0.80	−56.41
	2DV	1.12	11.52	8.66	155.982	7.32	66.00	−5.54	−90.66	−14.49	−215.77	−0.65	−61.61
	D(%)	6	6	5	10	1	9	0	9	5	13	−23	8
psw0p0	1DH	0.77	7.15	6.05	121.44	5.86	52.99	−2.59	−63.93	−7.02	−145.06	0	−34.06
	2DV	0.85	7.83	6.49	137.13	5.94	58.88	−2.49	−71.94	−7.70	−171.90	0.17	−39.21
	D(%)	9	9	7	11	1	10	−4	11	9	16	−	13
pbw0p5	1DH	0.73	6.38	5.55	116.46	5.51	51.01	−1.92	−59.56	−5.69	−135.91	0.14	−29.25
	2DV	0.80	7.06	5.99	132.10	5.59	56.86	−1.81	−67.40	−6.37	−162.61	0.29	−34.37
	D(%)	9	10	7	12	1	10	−6	12	11	16	−	15
ps1p5	1DH	1.10	9.89	8.53	176.62	8.41	77.35	−3.24	−91.51	−9.20	−207.85	0.12	−46.10
	2DV	1.20	10.89	9.16	199.94	8.52	86.07	−3.08	−103.32	−10.22	−247.85	0.35	−53.74
	D(%)	8	9	7	12	1	10	−5	11	10	16	−	14
pb0p5	1DH	0.75	7.09	6.00	120.866	5.83	52.84	−2.62	−63.92	−7.02	−144.46	0	−33.88
	2DV	0.82	7.75	6.42	136.38	5.89	58.63	−2.52	−71.88	−7.70	−171.17	−0.14	−38.98
	D(%)	9	9	7	11	1	10	−4	11	9	16	−	13

From Fig. 5.5, the transport of 2DV is averagely larger 5% than 1DH, except for 295a and 345a. When $f_{SUSW} = 0$, the ratio between 2DV and 1DH is more stable than $f_{SUSW} \neq 0$. For different wave conditions, high waves cause larger transport in 2DV. These differences also may be caused by hydrodynamic conditions due to different computational grids. In general, there are no significant differences between the sediment transports of both profile models. For the tide-averaged and cross-shore integrated transport on the profile, the magnitude of 2DV model is slightly larger than 1DH model, but for the total net longshore transport, most of the computed cases ($f_{SUSW} = 0$) in 2DV model have less magnitudes than 1DH model, about 91% ~ 94% of the latter. Even so, the differences between two profile models can be neglected for the sensitivity runs of this section.

5.2.3 Morphodynamic simulations of profile model

To evaluate the bathymetry changes due to different settings of transport factors, the morphodynamic simulations of profile model are carried out. The morphodynamic simulations of profile model still follow the procedure of Fig. 5.1. The simulations are also steered by Delft3D-FLOW and use the SZW model to perform wave computations. The morphological acceleration factor f_{MOR} and the simulation time for each wave boundary condition are the same as Table 5.1. Only the calibration cases with a northward net transport are used for morphodynamic simulations. Fig. 5.6 shows the final computed bathymetries of the 1DH (top plot) and 2DV (bottom plot) profile models.

In the figure, the final measured bathymetry may00 is still a bar-trough structure like the initial condition, but all the computed bathymetries nearly flatten the structure whether in 1DH or in 2DV model. The nourishment mostly keeps its original place, though its outline becomes smooth. The computed outcomes have good agreements with the final measured results at the seaward of the nourishment and have reasonable agreements at the swash zone (between -150 m and 0 m). However, both profile models can not reproduce the bar migration and the bar-trough structure.

The difference between the results of the calibration cases of each model is not so manifest that it is difficult to determine which one is the best. To assess the quality of each modelled result and give a quantitative criterion, the well-known Brier Skill Score (BSS) is used. The formula reads:

$$BSS = 1 - \frac{\langle (Y - X)^2 \rangle}{\langle (B - X)^2 \rangle} \tag{5.3}$$

where Y is computed bathymetry, X is measured final bathymetry, and B is baseline prediction. $\langle \cdots \rangle$ is averaging procedure over the cross section, i.e.,

$$\langle (Y - X)^2 \rangle = \frac{1}{n} \sum_{i=1}^{n} (y_i - x_i)^2 \tag{5.4}$$

$$\langle (B - X)^2 \rangle = \frac{1}{n} \sum_{i=1}^{n} (b_i - x_i)^2 \tag{5.5}$$

in which n is the number of bathymetry points on the profile.

The performance of a model relative to a baseline prediction can be judged by calculating the Brier Skill Score (BSS). This skill score compare the mean square difference between the prediction and observation with the mean square difference between baseline prediction and observation. Perfect agreement gives a BSS of 1, whereas modelling the baseline condition gives a score of 0. If the model prediction is further away from the final measured condition than the baseline prediction, the skill score maybe negative. The BSS is very suitable for the prediction of bed evolution. The baseline prediction for morphodynamic modelling will usually be that the initial bed remains unaltered. In other words, the initial bathymetry is used as the baseline prediction for the final bathymetry. A limitation of the BSS is that it cannot account for the migration direction of a bar; it just evaluates whether the computed bed level is closer to the measured bed level than the initial bed level. If the computed bar migration is in the wrong direction, but relatively small; this may result in a higher BSS compared to the situation with bar migration in the right direction, but much too large. The BSS will even be negative, if the bed profile in the latter situation is further away from the measured profile than the initial profile.

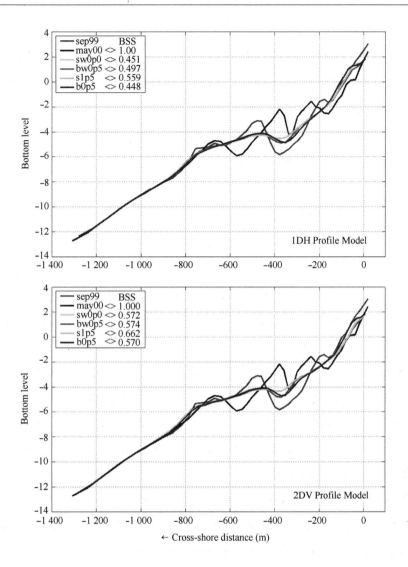

Fig. 5.6 Computed bathymetries of 1DH and 2DV profile models. "sep99" and "may00" are the measured bathymetries of September 1999 and May 2000. The former is the initial bottom of computation, and the latter is the final measured result. BSS is Brier Skill Score.

On this stage, the BSS is only used to rank the modelled results. A larger score means a better result. The BSSes for the calibration cases are shown in the legend of Fig. 5.6. For both models, the ranks (from the best to the worst) for the four cases are identical: ① ps1p5, ② pbw0p5, ③ psw0p0 and ④ pb0p5. For any profile model, the computed bathymetries in the seaside of the trough (offshore-450 m) almost overlap one another. The more evident differences between these cases appear in the trough (− 450 m ~ − 250 m) and in the swash zone (− 150 m ~ 0 m). The results of psw0p0 and pb0p5 almost overlap each other. There are deeper troughs in pbw0p5 and pb0p5 than in psw0p0 and ps1p5. In the swash zone, pbw0p5 and ps1p5 are closer to the final measured bottom than other two cases. Based on the BSS qualification, the case ps1p5 is thought as the best one for morphological simulation in both models.

5.2.3.1 Intercomparison on bottom change of both models

The differences between the bottom levels of both models (2DV minus 1DH) are shown in Fig. 5.7. Positive means the bottom level of 2DV is higher than 1DH, otherwise negative means 2DV is lower than 1DH. According to the figure, the bottom levels of 2DV are slight higher than 1DH offshore − 200 m, and the maximum is about 50 cm. At the landside of − 200 m, the bottoms of 2DV are lower than 1DH, the maximum difference is about − 60 cm. In the nourishment area, the bottoms of 2DV are slightly lower than 1DH. The larger differences between both models occur at the outer slope of the nourishment (− 800 m), at the trough of offshore bars (− 300 m), and at the mean sea level in the swash zone (− 100 m). These areas have a common characteristic: steeper slopes than other parts of the profile, which indicates that slope effect may play a role in numerical simulations, but the differential values are small in the present application.

Comparing the BSSes of both models, 2DV has higher scores, despite of both can not reproduce the bar migration. So the performance of 2DV is better than 1DH on the profile modelling of morphological evolution. Certainly, the differences between them are quite small.

5.2.4 Conclusions on profile modelling

The performances of both profile models show good compatibility between them. For different wave boundary conditions, the 2DV model has larger sediment transport than 1DH, but for the total net transport (integrating all the wave conditions), the value of 2DV is slightly less than 1DH. The

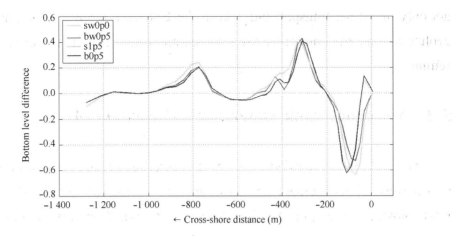

Fig. 5.7 Difference between modelled bathymetries of 1DH and 2DV profile models. Positive means the bottom level of 2DV is higher than 1DH, otherwise negative means 2DV is lower than 1DH.

modelled bathymetries of 2DV have higher BSSes than 1DH, while the differences between them are quite small. With respect to more computation efforts in 2DV model, 1DH profile model has faster computation effectivity and has not lost accuracy in present application. Based on the results of both models in this section, the 1DH profile model can substitute the 2DV model to carry out the calibration on transport factors for the Egm2004 cases.

The net transport is sensitive to the settings of transport factors. The f_{SUSW} is more important than other factors. When $f_{SUSW}=0$, the f_{BED} is less sensitive than other factors. In terms of the results of UNIBEST, the net transport with $f_{SUSW}=0$ and $f_{SUS}=1.5$ is closer to the predicted values.

There are no significant differences between the modelled final bathymetries of four sensitivity runs, but all the outcomes have large discrepancies to the final measured results. The bar-trough structure on the profile is almost flattened in the modelled results. All the sensitivity runs can not exactly reproduce the bar migration of the measured data. Intercomparing four sensitivity runs on morphological simulation, the case with $f_{SUSW}=0$ and $f_{SUS}=1.5$ has the highest BSS.

The calibrated factors of the cases psw0p0 and ps1p5 are finally chosen to extend to 3D area morphodynamic simulation. The latter ranks the first place not only in the net transport but also in the BSS assessment on morphological evolution, and the former is used for comparison since it has more default settings.

5.3 Morphodynamic simulations of area model

The calibration carried out by the profile models demonstrates the sensitivities of the related transport factors to the net transport and the morphological developments. The factors determined by the calibration are applied to 3D area model simulation in this section. Two fully 3D area modelling cases are performed in this section, which use the transport factors of the sensitivity runs psw0p0 and ps1p5. Now the names of these cases are modified to Egm2004-VR3D-sw and Egm2004-VR3D-su.

To compare with the previous morphodynamic modelling case Egm2002-Bk2DH (van Duin, 2002), two 2DH morphological simulations with "online transport approach" are also executed in this section. Their names are used as before, Egm2004-Bk2DH and Egm2004-VR2DH. Note the transport factors used in present Egm2004-VR2DH apply the calibrated values in the 1DH/2DV profile sensitivity run ps1p5. There are totally four (area model) morphodynamic simulations performed in this section. For clarity reason, the main differences between these cases are summarised in Table 5.7. The previous modelling case Egm2002-Bk2DH is also listed in the table for reference.

Table 5.7 Morphodynamic modelling cases

Name	Dim.	Formula	Transport factor/approach setting
Egm2002-Bk2DH	2DH	Bijker 1971	(van Duin, 2002), "offline"
Egm2004-Bk2DH	2DH	Bijker 1971	same setting as Egm2002, online
Egm2004-VR2DH	2DH	van Rijn 1993	calibrated by 1DH model, online
Egm2004-VR3D-sw	3D	van Rijn 1993	calibrated by 2DH model, online
Egm2004-VR3D-su	3D	van Rijn 1993	calibrated by 3DH model, online

The 3D area model setup is nearly the same as the 2DH area model, except the vertical layers which have only 1 layer in 2DH and 11 layers in 3D. The 2DH and 3D area models share the identical physical and numerical parameter settings. The morphological simulations are following the scenario discussed in the beginning of this chapter. The modelled results are described below on modelled bottoms, profile changes, volume changes, and longshore bar migrations.

5.3.1 Modelled bottoms

The top views (2D-plot) of measured bathymetries in September 1999 and May 2000 and the hindcasted bathymetry in the previous study Egm2002-Bk2DH are shown in Fig. 5.8. The total area of interest is subdivided into 20 boxes to interpret the volumes of sediment transport.

The shore parallel boundaries are chosen based on morphological development of the cross sections. The main aim is to keep moving bars within their section. At $x = -600$ m (x is the cross-shore coordinate) the cross sections are more or less stable therefore a boundary is chosen at the location. The nourishment remains seaward of this boundary. The next shore parallel boundary is chosen at $x = -300$ m, on the spot where initially the trough between outer and inner bar is located. The biggest part of the outer bar remains seaward of this boundary. For observation of the effects of the shoreface nourishment with regard to the beach, a boundary is chosen at $x = -100$ m, which corresponds with the NAP -1.00 m level. So the area consists of 4 longshore sections A, B, C, D.

The cross-shore section boundaries are also selected based on the principle that the longshore movement of bars should be kept as much as possible within their sections. Two boundaries are set north and south of the nourishment, at $y = +1\,500$ m and $y = -1\,500$ m (y is the longshore coordinate), the nourishment displacements do not cross these boundaries. The nourishment is split up in three sections, containing a centre part, a northern part and a southern part. So there are 5 cross-shore sections (1, 2, 3, 4, and 5) in the

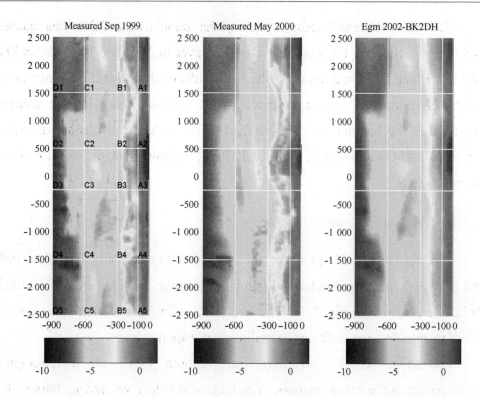

Fig. 5.8 Top views of measured bathymetries and computed bathymetry of Egm2002-Bk2DH. The results of Egm2002-Bk2DH are taken directly from the previous study.

area, which lead to 20 boxes.

The modelled bottoms are shown in Fig. 5.9, which includes Egm2004-Bk2DH, -VR2DH, -VR3D-sw, and -VR3D-su (These four cases are generally named as Egm2004-# in the following). From the figure, the outer bar and the inner bar are flattened, and the trough between the bars is filled in the Egm2004 modelled results. The bar-trough pattern which can be observed in the measured bathymetries and the results of Egm2002-Bk2DH in Fig. 5.8, disappears in the Egm2004-# results. The complex bathymetries in the surf zone and the swash zone (intersections of rip channels and swash bars, -300 m ~ -100 m) become smooth and uniform. The small headlands and bays formed in the measured results of May 2000. Such landforms can also be observed in the Egm2004-# modelled results, but they are not distinct in Egm2002-Bk2DH. The edges of such landforms in the modelled results are less curved than the measured results. The nourishment nearly remains its original place and shape in the final measured bathymetry and in all the modelled bottoms as well. More details of the bottom changes are discussed in the following subsections.

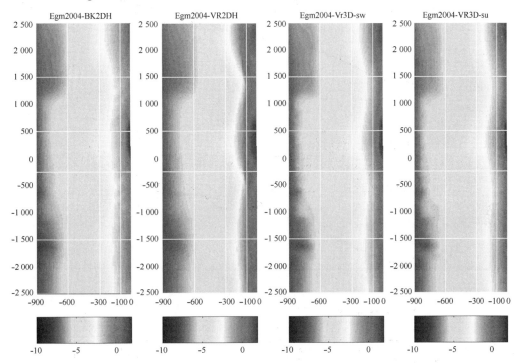

Fig. 5.9 Topviews of modelled bathymetries of Egm2004-Bk2DH, -VR2DH, -VR3D-sw, and-VR3D-su

5.3.2 Profile changes

To demonstrate the vertical changes of the modelled bottoms, the profile plots (side view) of the specified cross sections are given here. Fig. 5.10 shows the bottom development of three cross sections, north, middle, and south, which are located nearly at 1 500 m, 0 m, and -1 500 m of longshore coordinates in Fig. 5.8.

From the measured data, the outer bar moves onshore and the inner bar does offshore, which makes the trough between two longshore bars narrower. The middle section which cuts across the nourishment has the largest changes from the initial situation to the final measured results, and the north section

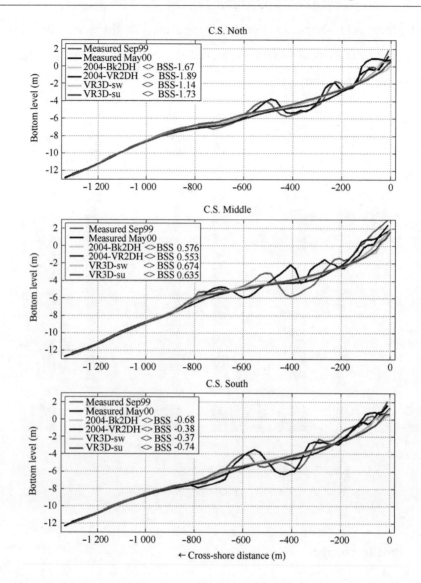

Fig. 5.10 Computed profiles of specified cross sections North, Middle, and South

has the smallest changes relatively. The natural bar-trough structure is flatten in all modelled results. Larger discrepancies between the modelled bottoms occur inside the surf zone. The nearshore swash bar is overestimated in Egm2004-VR2DH, however other cases are too flat to form an obvious swash bar. At the steeper slopes of the initial profile, i. e. the outer slopes of the nourishment and the inner bar, the bottom levels predicted by Egm2004-VR2DH are much lower than other cases. It is possible that Egm2004-VR2DH has stronger onshore transport than other Egm2004-# cases.

In general, there are no significant differences between all the modelled profiles. The BSS method is used again to give a quantitative criterion to assess the quality of modelled result. The BSSes of the modelled results which use the initial bathymetry (September 1999) as the baseline prediction are shown in the legend of Fig. 5.10. In all the three sections, the scores of Egm2004-VR3D-sw are higher than other cases. The skill scores of each case on the north section and the south section are negative, which means the modelled results are further away from the final measured condition than the initial condition. However, the scores of all modelled results get positive marks on the middle section, which is caused by the larger difference between the initial condition and the final measured condition around the nourishment. This point also indicates that the larger morphologaical evolutions take place in the nourished zone. Even so, Egm2004-VR3D-sw has the highest scores in each cross section, comparing with other simulation cases.

5.3.3 Volume changes

In this part, the volume changes of each box are calculated. The sedimentation and erosion volumes have been determined by subtraction of two different bathymetries. The plots can give a general idea of the amount of sedimentation and erosion and the location. Further, the sedimentation and erosion volumes in longshore sections and cross-shore sections are integrated.

Fig. 5.11 shows the sedimentation and erosion patterns of the measured data and all modelled results, in which the measured data of September 1999 is the initial bottom. According to the measured data, the main sediment deposit took place in the trough behind the nourishment, and the accretion area expended southward and northward about 500 m along the inner slope of the outer bar. At relatively shallow areas of the trough ($-2\,000$ m and $+2\,000$ m), significant sedimentations occurred against the outer slope of the inner bar. Main erosions happened on the outer bar and close to the nourishment. The front of the nourishment also suffered erosion, but in a weaker intensity. Minor sedimentation and erosion appeared here and there in the surf zone and

Shore Nourishment and Morphodynamic Modelling

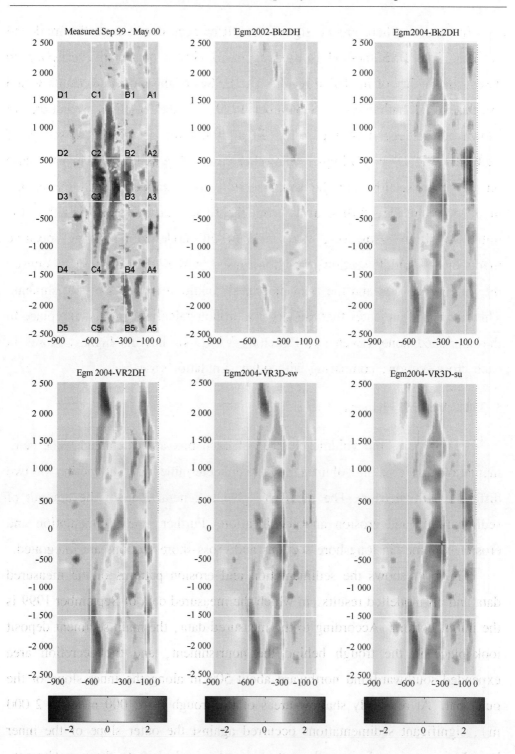

Fig. 5.11 Sedimentation and Erosion of measured and modelled bathymetries.
Egm2002-Bk2DH is taken directly from the previous study.

the swash zone.

In modelled results of Egm2002-Bk2DH shows little change. The erosions along the crests of the outer bar and the inner bar can be observed. The accretions at the shallow area of the trough also evident, though their intensities are still small. Dotted sedimentations appear on two sides of the inner bar. The results of Egm2004-# all present significant sedimentations in the trough and close to the outer slope of the inner bar. The sedimentation and erosion patterns show a distinct sedimentation strip along the original trough which looks like to combine the separate sedimentation patches in the measured data. Simultaneously large erosions take place on the outer bar and the inner bar, while the intensity on the latter is less than the former. In the swash zone, except that larger accretions along the shoreline appears in Egm2004-VR2DH, all other cases have erosions along the shoreline in different extents. Another apparent feature in all Egm2004-# cases is that there are sedimentations in the deep water to the south and the north of the nourishment, in spite of they are relatively slight in Egm2004-VR2DH. This indicates again that Egm2004-VR2DH makes more onshore transport than other cases. In a word, the modelled sedimentation and erosion patterns of Egm2004-# show a quite similar trend to the measured data, but locally there are large discrepancies in detail.

Although all the modelled results can not exactly reproduce the measured sedimentation and erosion pattern, Egm2004-# have more reasonable performances than Egm2002-Bk2DH. To provide a quantitative expression on the volume changes, the detailed magnitudes in the boxes of each modelling case are summarised in Table 5.8 in which the results of Egm2002-Bk2DH are taken directly from the previous study. The table not only gives the volume changes in each box, but also shows the volume changes in each longshore or cross-shore section. The quantities of volume change are also represented in Fig. 5.12. The volume changes now are expressed as the averaged sedimentation and erosion thickness (unit: m) in each box. Fig. 5.13 shows the integrated volume changes in longshore and cross-shore sections.

Table 5.8 Sediment volume changes in the modelled area from Sept. 1999 to May 2000 (Unit: m^3)

L. S. Box	A1	A2	A3	A4	A5	A6
Measured	-265	-28 744	-45 005	-49 085	5 713	-117 387
Egm2002-Bk2DH	-1 294	-1 837	-187	-514	787	-3 046
Egm2004-Bk2DH	-72 872	-76 022	-79 026	-98 944	-82 081	-408 945
Egm2004-VR2DH	37 482	59 334	14 899	74 458	61 556	247 728
Egm2004-VR3D-sw	1 165	-17 427	-37 737	-37 491	-23 520	-115 010
Egm2004-VR3D-su	-45 727	-61 853	-72 239	-96 633	-74 098	-350 550
L. S. Box	B1	B2	B3	B4	B5	B6
B7Measured	41 394	-24 238	100 463	-30 238	30 319	117 700
Egm2002-Bk2DH	13 634	7 490	12 737	18 173	-11 004	41 030
Egm2004-Bk2DH	-41 965	-32 201	-52 222	-40 641	-72 372	-239 403
Egm2004-VR2DH	76 923	-16 709	17 615	-37 809	-36 412	3 608
Egm2004-VR3D-sw	-41 267	-27 225	-25 353	-40 439	-68 730	-203 013
Egm2004-VR3D-su	-70 430	-32 110	-33 950	-41 753	-74 732	-252 975
L. S. Box	C1	C2	C3	C4	C5	C6
CMeasured	21972	89 258	43 330	49 895	11 858	216 314
Egm2002-Bk2DH	16 429	1 565	9 818	-27 171	8 492	9 133
Egm2004-Bk2DH	31240	139860	98923	171074	106298	547396
Egm2004-VR2DH	-84 424	51 037	49 954	76 131	27 793	120 491
Egm2004-VR3D-sw	-15 547	113 221	75 143	146 916	59 000	378 733
Egm2004-VR3D-su	1 975	158 546	89 126	212 502	88 849	550 997
L. S. Box	D1	D2	D3	D4	D5	SumD
Measured	-11 778	-11 309	-22 089	12 841	33 610	1 276
Egm2002-Bk2DH	10 612	17 384	-7 859	37 104	7 681	64 923
Egm2004-Bk2DH	54 190	11 920	-14 212	-249	44 407	96 056
Egm2004-VR2DH	5 334	-42 230	-77 904	-79 958	-17 284	-212 041
Egm2004-VR3D-sw	71 515	10 260	-29 359	-3 179	51 604	100 841
Egm2004-VR3D-su	118 167	37 844	-14 821	26 540	87 972	255 701
C. S. Section	1	2	3	4	5	Total
Measured	51 323	24 966	76 699	-16 586	81 501	217 903
Egm2002-Bk2DH	39 382	24 602	14 509	27 591	5 956	112 040
Egm2004-Bk2DH	-29 408	43 556	-46 538	31 241	-3 748	-4 896
Egm2004-VR2DH	35 315	51 432	4 564	32 822	35 653	159 785
Egm2004-VR3D-sw	15 867	78 829	-17 307	65 808	18 354	161 551
Egm2004-VR3D-su	3 985	102 427	-31 884	100 655	27 992	203 174

L. S. means longshore; C. S. is cross-shore.

Chapter 5 Morphodynamic Validation

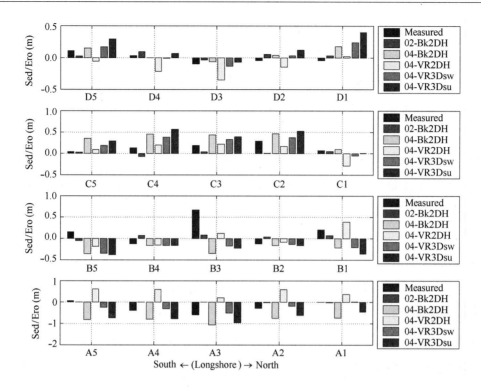

Fig. 5.12 Averaged sedimentation and erosion thickness of volume boxes for each case. Egm2002-Bk2DH is taken directly from the previous study. Positive means sedimentation and negative is erosion.

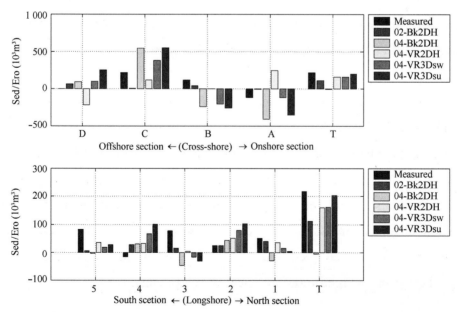

Fig. 5.13 Sedimentation and erosion volumes of cross-shore and longshore sections for each case. T means "total", i.e. the summation of all the sections. The T values of both plots are identical and in different scales. Positive means sedimentation and negative is erosion.

First to check the total volume change in the whole area, the measured data is 217 903 m³ sedimentation. The total change of Egm2004-Bk2DH is in a different trend other than the measured results, which is 4 896 m³ erosion and equivalent to-1 mm over the evaluated area 5 000 × 900 m². Other cases all show sedimentations in the area. The result of 2004-VR3D-su has the closest result to the measured data. The ratios of the modelled results to the measured data are 93% for Egm2004-VR3D-su, 74% for Egm2004-VR3D-sw, 73% for Egm2004-VR2DH, and 51% for Egm2002-Bk2DH.

Secondly, for cross-shore sections, the modelled results can reflect a reasonable trend of volume changes in Section D, C, and A, except for the case Egm2004-VR2DH. But in the longshore section B, most modelled results have erosions other than the measured sedimentation. The relatively complex bathymetries and hydrodynamic conditions in this section may cause the difficulties for modelling. Although the total volume of Egm2004-VR3D-su is closer to the measured results, its magnitudes in each longshore section are much larger than the measured volumes. The total volume changes of Egm2004-VR2DH and Egm2004-VR3D-sw are close to each other, but their changes in each section are quite different. The cases Egm2004-VR3D-sw and Egm2004-VR3D-su have good compatibility, since the only difference between them is the value of the factor f_{SUS}.

At last, for longshore sections. The measured data show that the sedimentations mainly go to Section 5 and Section 3, which take 37% and 35% of the total sedimentation volume. The sedimentation in Section 2 is about half of Section 1, while a small volume is eroded in Section 4. However, all the modelled results show sedimentations in Section 4 and Section 2. The modelled erosions almost take place in Section 3. So, the total volume changes in area are well predicted, but the model performances are poor for the prediction on detailed morphological developments.

5.3.4 Longshore bar migrations

According to the final measured bottom in Fig. 5.8, the bar-trough structure was still existed. The outer bar migrated onshore due to the redistribution of filled and trapped sand in the area. At the same time, the outline of the inner bar became more curved than in the initial condition. In the swash zone, there still presented complex bathymetries with swash bars and rip channels.

In Egm2002-Bk2DH, the modelled sedimentation and erosion were quite small, so no significant bottom evolutions were formed. The outer bar almost remained its original place, and the trough also did not change its outline much. The bar-trough structure was not evidently effected. The longshore bathymetries in the swash zone became uniform, and the outline of the swash bar seemed obscure. The bar migration could not be represented in the modelling.

In Egm2004-# cases, the offshore bars completely disappear, and the trough is filled up. The bar-trough structure is flattened. In the swash zone the simple bathymetry replaces the swash bars and rip channels in measured results. Although the sedimentation and erosion pattern is quite reasonable, but the longshore bar migration is not exactly realised in these modelling cases.

5.3.5 Discussions on area morphodynamic modelling

There are many differences between Egm2002-Bk2DH and Egm2004-Bk2DH, though both were run in 2DH and used the same transport formula with identical settings. The differences are the computational grids, the types of tide boundary conditions, and implementations of morphodynamic simulations, i.e. "offline" or "online" transport approach. For the total volume changes, Egm2002-Bk2DH was 51% of the measured sedimenta-tion, while the latter gets averaged 1 mm erosions in the area. For modelled bottom, the former did not change it significantly and remained the bar-trough structure, but the latter flattened the bottom. It is remarkable that the transport settings were not calibrated in Egm2004-Bk2DH, and totally different approaches were used in both model ("offline" versus "online"), which may play a key role in causing the large discrepancy between their results.

According to the analyses of modelled results, each case of Egm2004-# has similar sedimentation and erosion patterns. The modelled bottoms of Egm2004-# cases also have similar appearances. The main differences between these cases are the transport formula and the dimensions of computa-tional grids. Egm2004-Bk2DH uses *Bijker* 1971 formula, while other three cases all use *van Rijn* 1993 formula. Now we focus on the three cases with *van Rijn* 1993 formula. All these cases have good results to predict the sediment budget. The performances of Egm2004-VR3D-sw and Egm2004-VR3D-su are much the same on profile evolutions and on volume changes in each box and therefore longshore and cross-shore sections, since their difference is only the factor f_{SUS} after all. Contrasted to these two 3D cases, Egm2004-VR2DH shows stronger onshore transport, which causes larger discrepancies in cross-shore sedimentation and erosion distribution, see Fig. 5.13.

The area modelling cases with *van Rijn* 1993 formula appear good performances to predict the total volume changes, and quite reasonable results to model morphological evolutions. More detailed calibrations may result in better outcomes for large-scale sand budget prediction, even for local morphological developments.

5.4 Comparison of profile and area modelling

Except Egm2004-Bk2DH, other area modelling cases apply van Rijn 1993 formula and use the transport factors calibrated by the 1DH/2DV profile models. Before the area morphological modelling is performed, an assumption was made: the sediment transport and the morphological evolution of profile model represent the processes of corresponding area model. Now the results of the profile models and the corresponding area models can be used to test the assumption. The only common outcomes of both types of models are the final simulated profiles, i.e. the bottoms of the middle cross section (Jan van Speijk), see Fig. 5.14.

Chapter 5 Morphodynamic Validation

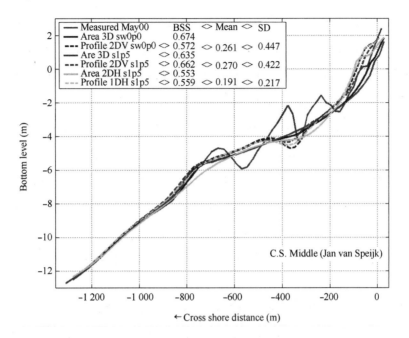

Fig. 5.14 Comparison of computed profiles of profile model and area model. "Mean" and "SD" are the mean and the standard deviation of the bottom level difference between profile model and area model. The solid lines indicate the area models, and the dash lines do the corresponding profile models.

From the figure, whether the profile models or the area models flatten the bottom, but small troughs can be observed evidently in the profile models. The area models do not exactly duplicate the results of the profile models. In general, the modelled bottom levels of the profile models are higher than the area models, especially in the swash zone (-200 m ~ 0 m). The possible explain is that more erosion due to longshore transport gradients presents near beach in area models. The BSSes of all the modelled results are shown in the legend of the figure. There is no instinct relationship between the BSSes. The area model Egm2004-VR3D-sw has a higher score than its corresponding profile model, while Egm2004-VR3D-su has a lower score, and Egm2004-VR2DH almost has the same score as the profile model.

Standard deviation (SD) is used to evaluate the differences of the computed bottoms of both models. The means and the SDs of the differences are also shown in the legend of the figure. From the figure, Egm2004-VR2DH has the smallest mean, and the smallest standard deviation, which indicates Egm2004-VR2DH has better compatibility with its corresponding profile model than other area models. The means of the relative differences between two 3D models and their corresponding profile models are 0.261 m for Egm2004-VR3D-sw and 0.270 m for Egm2004-VR3D-su, and the standard deviations are 0.447 m, 0.422 m respectively.

The profile models and the area models have quite good agreements on the final computed bottoms except near the beach. The results prove that profile model not only can be used to calibrate the modelling factors for corresponding area models, but also can be used to predict the morphological evolutions.

In this study, the middle cross section (Jan van Speijk) is chosen as the representative section to perform profile modelling, since the nourishment is the focused area of interest. In fact, the calibrated transport factors by the profile model are finally used in all computational grid cells of the area model. Which cross section can represent all the cross sections? Although the performances of the profile models in present study show quite good agreements with the area modelling, but it would give much better results if choosing a more representative profile for profile modelling. In van Rijn *et al.* (2003), a procedure of choosing representative profile is outlined. However, for the profile modelling based on Delft3D, how to choose or schematise the representative profile is still a valuable issue for future study.

5.5 Conclusions

The morphological evolution of the Egmond coastal area concerns the fully coupled activities of waves, flow, and sediment transport and bed level variations. This fully coupled dynamic system is modelled in this study based on Delft3D-FLOW. The morphodynamic simulations follow a prescribed scenario, in which twelve wave conditions are included. The morphological developments are simulated one by one for all the wave conditions. The

Delft3D-FLOW module steers the simulation procedure, by distributing the simulation time and the morphological acceleration factor for the flow computation coupled with each wave condition.

The modelled Sediment transport is sensitive to the settings of the transport factors used in *van Rijn* 1993 formula. Such factors must be calibrated before final area morphodynamic simulations. The profile models are developed to fulfil the calibration against the results of UNIBEST, due to the considerable time efforts of full 2DH and 3D area modelling.

The profile models of 1DH and 2DV correspond respectively to 2DH and 3D area models. The computed results of both profile modelling show good compatibility. The differences of sediment transport and morphological evolutions between two models are quite small. It is concluded that the 1DH profile model can substitute the 2DV profile model in the present case, since the former has faster computation effectivity without accuracy lost.

The results of the profile modelling indicate that the net transport is more sensitive to the factor f_{SUSW}. With respect to the results of UNIBEST, the net transport with $f_{SUSW} = 0$ and $f_{SUS} = 1.5$ is closer to the predicted values. The profile models can not exactly reproduce the final measured bottom. The modelled bottom is flattened and the bar-trough structure nearly disappears.

Two 2DH and two 3D area modelling cases are carried out in this chapter. Egm2004-Bk2DH used the same transport formula (*Bijker* 1971) and factor settings as the previous study, and other cases used van Rijn 1993 formula and the transport factors calibrated by the profile models. All the cases have similar appearances in modelled bottoms and in sedimentation and erosion patterns. Their performances are more reasonable than the previous study Egm2002-Bk2DH. The cases with *van Rijn* 1993 formula well predict the volume changes in the area, while Egm2004-Bk2DH does not give a good prediction which may be due to that its transport settings are not calibrated. Although all the cases show reasonable results on morphological evolutions, the locally detailed morphological features are not yet well modelled.

For the modelling cases with the same transport formula van Rijn 1993, their results are consistent. Two 3D case are much the same on morphological developments and volume changes. The 2DH case shows stronger onshore transport than the 3D cases.

The profile models and the area models have quite good agreements on the final computed bottoms. The results prove that profile model not only can be used to calibrate the transport factors for corresponding area models, but also can be used to predict the morphological evolutions.

Chapter 6

Conclusions and Recommendations

6.1 Conclusions

The main objectives of this study were to validate the hydrodynamic and morphodynamic Delft3D modelling on the Egmond shoreface nourishment. With respect to the implementation of the modelling, the study was focused on two aspects: hydrodynamic modelling and morphodynamic modelling.

6.1.1 Egmond Shoreface Nourishment

At Egmond aan Zee a shoreface nourishment has been applied in 1999. The nourishment is approximately 2 km long and 200 m wide. The total sand volume is 900 000 m^3 with the characteristic volume 400 m^3/m. To study the influence of the shoreface nourishment area on the morphodynamics, the analysis of bathymetry data of the nourishment area indicate:

(1) During the first 2 years (May 1999 to April 2001) the shoreface nourishment hardly changed in height or location and, therefore, did not contribute to the beach sand volume directly, i. e., by redistribution of the nourished sand.

(2) The inner and outer bars, on the other hand, showed a large shoreward migration and a trough was generated between the outer bar and the shoreface nourishment area. The shoreface nourishment area seemed to act as the new outer bar, taking over the function of the original outer bar. The

survey of April 2002 showed an end to this trend. The system seemed to return to its natural three-bar system.

(3) The shoreface nourishment was expected to diffuse, but surveyed data showed that the shoreface nourishment did not diffuse in the first two years. The scale (amplitude and length) of the shoreface nourishment was probably too large as a result of which it did not diffuse and morphodynamic interaction occurred with the autonomous system. After a period of two years, the shoreface nourishment started to diffuse resulting in a lower amplitude and shoreward movement of the outer bar.

(4) Accretion occurred shoreward of the shoreface nourishment, indicating that the shoreface nourishment functions as a reef with a lee-side effect shoreward of the nourishment area. The total area shows a net gain of sand volume during the overall period (May 1999 to April 2002), including placement of the shoreface nourishment and beach nourishment. After three years, about 45% of the nourishment is still present.

6.1.2 Hydrodynamic modelling

The aims of hydrodynamic modelling were to setup an effective Delft3D model, and to provide reliable hydrodynamic results for further morphodynamic simulations. The main conclusions in hydrodynamic modelling are summarized below:

(1) The present Egmond 3D model consists of two grids used respectively for the FLOW and WAVE modules. The flow/morphological grid covers the area of $1\,300 \times 5\,200$ m^2 which is nested within the wave grid of $2\,400 \times 14\,900$ m^2. The vertical profile of the flow grid is separated into 11 layers for 3D simulation. The bottoms of the grids use the measured bathymetry dated on 01 September 1999.

(2) The schematized tide adopts the result of previous study. However the boundary conditions have to be regenerated to suit the new grid. Riemann variants are derived as the lateral boundary conditions of the model. The calibration against the previous study shows that the new model not only well

reproduces the tidal currents but makes more stable results as well.

(3) The wave computation of the model is based on the default settings of the SWAN system. The boundary conditions (the schematized wave conditions) are also copied directly from the previous study. The model can reflect the offshore wave propagation over the area of interest. The output of the wave computation provides an orderly wave field for the flow grid. Most of wave breaking happens on the longshore bars and the nourishment. The wave energy dissipation mainly concentrates on the longshore bars. Wave-current interactions significantly change the flow pattern within the surf zone.

(4) The formula of sediment transport used in the model is van Rijn's, which includes the transport caused by both currents and waves. The local sediment transport is sensitive to the wave boundary conditions. Moreover, the computed transport relies on the settings of the transport factors used in the formula. With the default settings of transport factors, there are larger discrepancies between the present study and the previous studies on the net transport. The reasons may be caused by the different formulas adopted in the models and the related settings of the transport factors.

6.1.3 Morphodynamic modelling

The morphodynamic modelling was based on the consequences of hydrodynamic modelling. The performances of the morphodynamic simulations were analyzed against the measured data. The major findings are:

(1) The morphological evolution of the Egmond coastal area concerns the fully coupled activities of waves, flow, and sediment transport and bed level variations. This fully coupled dynamic system is modelled in this study based on Delft3D-FLOW. The morphodynamic simulations follow a prescribed scenario, in which twelve wave conditions are included. The morphological developments are simulated one by one for all the wave conditions. The FLOW module steers the simulation procedure, by distributing the simulation time and the morphological acceleration factor for the flow computation coupled with each wave condition.

(2) The modelled Sediment transport is sensitive to the settings of the transport factors used in *van Rijn* 1993 formula. Such factors must be calibrated before final area morphodynamic simulations. The profile models are developed to fulfil the calibration against the results of UNIBEST, due to the considerable time efforts of full 2DH and 3D area modelling.

(3) The profile models of 1DH and 2DV correspond to 2DH and 3D area models. The results of profile modelling show good compatibility between both models. The differences of sediment transport and morphological evolutions between too models are quite small. It is concluded that the 1DH profile model can substitute the 2DV profile model for the present case Egm2004, since the former has faster computation effectivity and has not lost accuracy.

(4) The results of the profile modelling indicate that the net transport is more sensitive to the factor f_{SUSW}. With respect to the results of UNIBEST, the net transport with $f_{\text{SUSW}} = 0$ and $f_{\text{SUS}} = 1.5$ is closer to the predicted values. The profile models can not exactly reproduce the final measured bottom. The modelled result is flattened and the bar-trough structure nearly disappears.

(5) Four area morphodynamic modelling cases are performed. Egm2004-Bk2DH used the same transport formula *Bijker* 1971 and factor settings as the previous study, and other cases used *van Rijn* 1993 formula and the transport factors calibrated by the profile models. All the cases have similar appearances in modelled bottoms and in sedimentation and erosion patterns. They have more reasonable performances than the previous study. In quantitative volume changes, the cases with *van Rijn* 1993 formula are well predicted, while Egm2004-Bk2DH does not give a good prediction because its transport settings are not calibrated. All the cases show reasonable results on morphological evolutions, but the locally detailed morphological features are not well modelled. The performances of the area models are consistent with the results of Delft3D profile models.

(6) For the modelling cases with the same transport formula *van Rijn* 1993, their results have good compatibility. Two 3D case are much the same

on morphological developments and volume changes. The 2DH case shows stronger onshore transport than the 3D cases.

(7) The profile models and the area models have quite good agreements on the final computed bottoms. The results prove that profile model not only can be used to calibrate the modelling factors for the corresponding area models, but also can be used to predict the cross-shore morphological evolutions.

6.2 Recommendations for future study

On the basis of the modelling results of hydrodynamic and morphodynamic simulations in this study, the following recommendations are made:

(1) The net transport is dependent closely to the transport factor settings and local hydro-morphodynamic conditions. It is suggested to study on the reliabilities between the factors and local hydro-morphodynamic conditions. More detailed calibrations are suggested to test on the reliability between the transport factors and local hydrodynamic and morphodynamic conditions.

(2) The profile model deploys Surf Zone Wave model (roller with Snell's) to fulfil wave simulations, but in the present study the roller model was not calibrated. If data available, the calibrated profile model would draw better results.

(3) Profile model is a simplified area model, in which only one cross section is considered. How to choose or schematize this representative cross section could affect the final results. Sensitivity simulations addressing this can give further insight in the performance of the Delft3D models in profile mode.

(4) Profile model is very useful to fulfil some functions of area model, while the compatibility between them is recommended to be further verified in details.

(5) Bottom roughness in this study uses instant values. In reality, the bed roughness highly varies in time and in space. A roughness prediction has been developed in (van Rijn, 2003), which should be incorporated in future

modelling study.

(6) Further studies to determine the basic cause of bar-trough flattening are required. The causes could be bottom slope effects, phase lags related peak of transport and bar crest.

Bibliography

Ahrens, J. P. , Hands, E. B. 1998. Parameterizing beach erosion/accretion conditions. Proceedings of the 26th International Conference on Coastal Engineering, Copenhagen, Denmark: 2382 – 2394.

Bailard, J. A. 1981. An energetics total load sediment transport model for a plane sloping beach. *Journal of Geophysical Research: Oceans* (1978 – 2012), 86: 10938 – 10954.

Battjes, J. A. , Janssen, J. P. F. M. 1978. Energy loss and set-up due to breaking of random waves, *Proceedings* 16*th International Conference on Coastal Engineering*: 569 – 587.

Browder, A. E. , Dean, R. G. , Chen, R. 2000. Performance of submerged breakwater for shore protection. *Proceedings of the* 27*th International Conference on Coastal Engineering*: 2312 – 2323.

Bijker, E. W. 1971. Longshore transport computations, *Journal of Waterways, Harbours and Coastal Engineering Division*, 97: 687 – 701.

Booij, N. , Ris, R. C. , Holthuijsen, L. H. 1999. A third generation wave model for coastal regions: 1. Model description and validation, *Journal of Geophysical Research*, 104: 7649 – 7666.

Capobianco, M. , Hanson, H. , Larson, M. , et al. , 2002. Nourishment design and evaluation: applicability of model concepts, *Coastal Engineering*, 47: 113 – 135.

Davies, A. G. , van Rijn, L. C. , Damgaard, J. S. , et al. , 2002. Intercomparison of research and practical sand transport models, *Coastal Engineering*, 46: 1 – 23.

De Vriend, H. J. , Zyserman, J. , Nicholson, J. , et al. , 1993a. Medium-term 2DH coastal area modelling. *Coastal Engineering*, 21: 193 – 224.

De Vriend, H. J. , Capobianco, M. , Latteux, B. , et al. , 1993b. Long-term modelling of coastal morphology: a review. *Coastal Engineering*, 21: 225 – 269.

Elias, E. P. L. , Walstra, D. J. R. , Roelvink, J. A. , et al. , 2000. Hydrodynamic validation of Delft3D with field measurements at Egmond, *Proceedings* 27*th International Conference on Coastal Engineering*: 2714 – 2727.

Fredsøe, J. 1984. Turbulent boundary layer in wave-current interaction, *Journal of Hydraulic Engineering*, ASCE, 110: 1103 – 1120.

González, M. , Medina, R. , Losada, M. 2010. On the design of beach nourishment projects using static equilibrium concepts: Application to the Spanish coast, *Coastal Engineering*, 57: 227 – 240.

Grasmeijer, B. T. 2002. *Process-based cross-shore modelling of barred beaches*, PhD thesis, The Royal

Dutch Geographical Society/Faculty of Geographical Sciences, Utrecht University, The Netherlands.

Grasmeijer, B. T., van Rijn, L. C. 1998. Breaker bar formation and migration. *Coastal Engineering Proceedings*, 1 (26).

Grasmeijer, B., Walstra, D. J. R. 2003. Coastal profile modelling: possibilities and limitations, *Proceedings of Coastal Sediment* 2003, pp. 125 – 131.

Grunnet, N. 2002. *Post-Nourishment Volumetric Development at Terschelling*. Department of Physical Geography, University of Utrecht, Utrecht, the Netherlands.

Grunnet, N. M., Walstra, D. J. R., Ruessink, B. G. 2004. Process-based modelling of a shoreface nourishment, *Coastal Engineering*, 51: 581 – 607.

Hamm, L., Capobianco, M., Dette, H. H., et al., 2002. A summary of European experience with shore nourishment, *Coastal Engineering*, 47: 237 – 264.

Hanson, H., Brampton, A., Capobianco, M., et al., 2002. Beach nourishment projects, practices, and objectives-a European overview, *Coastal Engineering*, 47: 81 – 111.

Hoefel, F., Elgar, S. 2003. Wave-induced Sediment Transport and Sandbar Migration, *Science*, 299: 1885 – 1887.

Hoekstra, P., Houwman, K. T., Kroon, A., et al., 1996. Morphological development of the Terschelling shoreface nourishment in response to hydrodynamic and sediment transport processes. *Proceedings of the* 25*th International Conference on Coastal Engineering*: 2897 – 2910.

Holthuijsen, L. H. 2010. *Waves in Oceanic and Coastal Waters*, Cambridge University Press.

Holthuijsen, L. H., Booij, N., Herbers, T. H. C. 1989. A prediction model for stationary, short-crest waves in shallow water with ambient currents, *Journal of Coastal Engineering*, 13: 23 – 54.

Isobe, M., Horikawa, K. 1982. Study on water particle velocities of shoaling and breaking waves. *Coastal Engineering in Japan*, 25: 109 – 123.

Klein, M. D., Elias, E. P. L., Walstra, D. J. R., et al., 2001. *The Egmond model: Hydrodynamic validation of Delft3D with field measurements of Egmond-Main experiment October-November* 1998, Delft Hydraulics Report Z2394, Delft, the Netherlands.

Lamberti, A., Mancinelli, A. 1996. Italian experience on submerged barriers as beach defence structures. *Proceedings of the* 25*th International Conference on Coastal Engineering*: 2352 – 2365.

Lesser, G. R., Roelvink, J. A., van Kester, et al., 2004. Development and validation of a three-dimensional morphological model, *Coastal Engineering*, 51: 883 – 915.

Longuet-Higgins, M. S., Stewart, R. W. 1964. Radiation stresses in water waves; a physical discussion, with applications, *Deep sea Research and Oceanographic*, 11: 529 – 562.

Reniers, A. J. H. M., Roelvink, J. A., Thornton, E. B. 2003. Morphodynamic modelling of an embayed beach under wave group forcing, *Journal of Geophysical Research*, 108: X1 – 22.

Rienecker, M. M., Fenton, J. D. 1981. A Fourier approximation method for steady water waves. *Journal of Fluid Mechanics*, 104: 119 – 137.

Ris, R. C., Holthuijsen, L. H., Booij, N. 1999. A third generation wave model for coastal regions: 2.

Bibliography

Verification, *Journal of Geophysical Research*, 104, 7667 – 7681.

Roelvink, D., Reniers, A. J. H. M. 2012. *A guide to modeling coastal morphology*. World Scientific.

Roelvink, J. A., Stive, M. J. F. 1989. Bar-generating cross-shore flow mechanisms on a beach. *Journal of Geophysical Research: Oceans* (1978-2012), 94: 4785 – 4800.

Roelvink, J. A., Walstra, D. J. R. 2004. Keeping it simple by using complex models, Advances in Hydro-science and Engineering 6 (Proceedings 6th International Conference on Hydro-science and Engineering): 1 – 11.

Southgate, H. N. 1995. The effect of wave chronology on medium and long term coastal morphology, *Coastal Engineering*, 26, pp. 251 – 270.

Spanhoff, R., Biegel, E. J., van de Graaff, J., et al., 1997. Shoreface nourishment at Terschelling, The Netherlands: feeder berm or breaker berm? *Coastal Dynamics* ' 97, ASCE: 863 – 872.

Stive, M. J. F. 1986. A model for cross-shore sediment transport. *Coastal Engineering Proceedings*, 1 (20).

Sun, B. 2004. Validation of hydrodynamic and morphodynamic modelling on a shoreface nourishment at Egmond, the Netherlands, Delft Hydraulics Report Z3624.

van Duin, M. J. P. 2002. Evaluation of the Egmond shoreface nourishment: Part 3 Validation morphological modelling Delft3D-MOR, Delft Hydraulics Report Z3054/Z3148, Delft, the Netherlands.

van Duin, M. J. P., Wiersma, N. R. 2002. Evaluation of the Egmond shoreface nourishment: Part 1 Data analysis, Delft Hydraulics Report Z3054/Z3148, Delft, the Netherlands.

van Duin, M. J. P., Wiersma, N. R., Walstra, D. J. R., et al., 2004. Nourishing the shoreface: observations and hindcasting of the Egmond case, The Netherlands, *Coastal Engineering*, 51: 813 – 837.

van Rijn, L. C. 1993. *Principles of sediment transport in rivers, estuaries and coastal seas*, Aqua Publications, Amsterdam, the Netherlands.

van Rijn, L. C. 1995. Sand budget and coastline changes of the central coast of Holland between Den Helder and Hoek van Holland period 1964 – 2040. Delft Hydraulics Report H2129, Delft, the Netherlands.

van Rijn, L. C. 2000. General view on sand transport by currents and waves: Data analysis and engineering modelling for uniform and graded sand (TRANSPOR2000 and CROSMOR2000 models). Delft Hydraulics Report Z2899.20/Z2099.30/Z2824.30, Delft, the Netherlands.

van Rijn, L. C., Walstra, D. J. R., Grasmeijer, B., et al., 2003. The predictability of cross-shore bed evolution of sandy beaches at the time scale of storms and seasons using process-based Profile models, *Coastal Engineering*, 47: 295 – 327.

van Rijn, L. C., Walstra, D. J. R., Grasmeijer, B. T., et al., 2001. Hydrodynamics and morphodynamics in the surf zone of a dissipative beach, *Proceedings 4th Coastal Dynamics Conference*: 373 – 382.

Walstra, D. J. R. 2001. *Evaluation of UNIBEST-TC model.* Delft Hydraulics Report Z3148.10, Delft, the Netherlands.

Walstra, D. J. R., Roelvink, J. A., Groeneweg, J. 2001. Calculation of wave-driven currents in a 3D mean flow model, *Coastal Engineering* 2000, ASCE: 1051 – 1063.

Wiersma, N. R. 2002. Evaluation of the Egmond shoreface nourishment: Part 2 Validation morphological model UNIBEST-TC, Delft Hydraulics Report Z3054/Z3148, Delft, the Netherlands.

Work, P. A., Dean, R. G. 1995. Assessment and prediction of beach nourishment evolution. *Journal of Waterway, Port, Coastal, and Ocean Engineering*, 121: 182 – 189.

Work, P. A., Otay, E. N. 1996. Influence of nearshore berm on beach nourishment. *Proceedings of the 25th International Conference on Coastal Engineering*: 3722 – 3735.

Appendix

A. Bijker transport formula with wave effect (1971)

The Bijker formula sediment transport relation is a popular formula which is often used in coastal areas. It is robust and generally produces sediment transport of the right order of magnitude under the combined action of currents and waves. Bed load and suspended load are treated separately. The near-bed sediment transport (S_b) and the suspended sediment transport (S_s) are given by the formulations in the first sub-section. It is possible to include sediment transport in the wave direction due to wave asymmetry and bed slope, which are included:

$$\vec{S}_b = \vec{S}_{b0} + \vec{S}_{b,asymm} + \vec{S}_{b,slope} + \vec{S}_{s,slope} \qquad (A.1)$$

$$\vec{S}_s = \vec{S}_{s0} \qquad (A.2)$$

where S_{b0} and S_{s0} are the sediment transport in flow direction as computed according to the formulations of Bijker shown in below, and the asymmetry and bed slope components for bedload and suspended transport are followed. Both bedload and suspended load terms are incorporated in the bed-load transport for further processing. The transport vectors are imposed as bed-load transport vector due to currents S_{bc} and suspended load transport mag-nitude S_s, from which the equilibrium concentration is derived, respectively.

The basic formulation of the sediment transport formula according to Bijker is given by:

$$S_b = bD_{50} \frac{q}{C} \sqrt{g}(1 - \varepsilon)\exp(A_r) \qquad (A.3)$$

$$S_s = 1.83 S_b \left(I_1 \log_e \left(\frac{33.0h}{r_c} \right) + I_2 \right) \quad (A.4)$$

where, C is Chézy coefficient (as specified in input of Flow module), h is water depth, q is flow velocity magnitude, ε is porosity; and other coefficients:

$$A_r = \max[-50, \min(100, A_{ra})] \quad (A.5)$$

$$b = BD + \max\left[0, \min\left(1, \frac{(h_w/h) - C_d}{C_s - C_d}\right)\right](BS - BD) \quad (A.6)$$

$$I_1 = 0.216 \frac{\left(\frac{r_c}{h}\right)^{z_*-1}}{\left(1 - \frac{r_c}{h}\right)^{z_*}} \int_{r/h}^{1} \left(\frac{1-y}{y}\right)^{z_*} dy \quad (A.7)$$

$$I_2 = 0.216 \frac{\left(\frac{r_c}{h}\right)^{z_*-1}}{\left(1 - \frac{r_c}{h}\right)^{z_*}} \int_{r/h}^{1} \log_e y \left(\frac{1-y}{y}\right)^{z_*} dy \quad (A.8)$$

where, BS is coefficient b for shallow water (default value 5), BD is coefficient b for deep water (default value 2), C_s is shallow water criterion (H_s/h, default value 0.05), C_d is deep water criterion (default value 0.4), r_c is roughness height for currents (m).

$$A_{ra} = \frac{-0.27 \Delta D_{50} C^2}{\mu q^2 \left[1 + 0.5 \left(\psi \frac{U_b}{q}\right)^2\right]} \quad (A.9)$$

$$\mu = \left(\frac{C}{18 \log_{10}(12h/D_{90})}\right)^{1.5} \quad (A.10)$$

$$Z_* = \frac{w}{\frac{kq\sqrt{g}}{C} \sqrt{1 + 0.5 \left(\psi \frac{U_b}{q}\right)^2}} \quad (A.11)$$

$$U_b = \frac{\omega h_w}{2\sinh(k_w h)}, \quad \omega = \frac{2\pi}{T} \quad (A.12)$$

$$f_w = \exp\left(-5.977 + \frac{5.123}{a_0^{0.194}}\right), \quad a_0 = \max\left(2, \frac{U_b}{\omega r_c}\right) \quad (A.13)$$

$$\psi = C \sqrt{\frac{f_w}{2g}}, \text{ if wave effects included } (T>0), \text{ otherwise } 0 \quad (A.14)$$

where, C is Chézy coefficient (as specified in input of FLOW module), h_w is wave height (H_{rms}), k_w is wave number, T is wave period computed by the waves model or specified by user as T_{user}, U_b is wave velocity, w is sediment fall velocity (m/s), Δ is relative density $(\rho_s - \rho_w)/\rho_w$, κ is von Kármán constant (0.41).

The following formula specific parameters have to be specified in the input files of the transport settings: BS, BD, C_s, C_d, dummy argument, r_c, w, ε and T_{user}. If the Bijker formula is selected it is possible to include sediment transport in the wave direction due to wave asymmetry following the Bailard approach (Bailard, 1981; Stive, 1986).

Separate expressions for the wave asymmetry and bed slope components are included for both bedload and suspended load. Both extra bedload and suspended load transport vectors are added to the bedload transport as computed in the previous Equation (A.1):

$$\vec{S}_b = \vec{S}_{b0} + \vec{S}_{b,asymm} + \vec{S}_{s,asymm} + \vec{S}_{b,slope} + \vec{S}_{s,slope} \quad (A.15)$$

where the asymmetry components for respectively the bed load and suspended transport in wave direction are written as:

$$S_{b;asymn}(t) = \frac{\rho c_f \varepsilon_b}{(\rho_s - \rho)g(1-p)\tan\varphi}[|u(t)|^2 u(t)] \quad (A.16)$$

$$S_{s;asymn}(t) = \frac{\rho c_f \varepsilon_b}{(\rho_s - \rho)g(1-p)w}[|u(t)|^3 u(t)] \quad (A.17)$$

from which the components in ξ and η direction are obtained by multiplying with the cosine and sine of the wave angle θ_w and the bed slope components as:

$$S_{b;slope,\xi}(t) = \frac{\rho c_f \varepsilon_b}{(\rho_s - \rho)g(1-p)\tan\varphi}\left[\frac{1}{\tan\varphi}|u(t)|^3\right]\frac{\partial z_b}{\partial \xi} \quad (A.18)$$

$$S_{s;slope,\xi}(t) = \frac{\rho c_f \varepsilon_b}{(\rho_s - \rho)g(1-p)w}\left[\frac{\varepsilon_s}{w}|u(t)|^5\right]\frac{\partial z_b}{\partial \xi} \quad (A.19)$$

and similar for the η direction, where: $u(t)$ is near bed velocity signal (m/s), ρ is density of water (kg/m^3), ρ_s is density of the sediment (kg/m^3), c_f is coefficient of the bottom shear stress (constant value of 0.005), p is porosity (constant value of 0.4), φ is natural angle of repose (constant value of $\tan\varphi$

= 0.63), w is sediment fall velocity (m/s), ε_b is efficiency factor of bed load transport (constant value of 0.10), ε_s is efficiency factor of suspended transport (constant value of 0.02, but in implemented expression for suspended bed slope transport the second ε_s is replaced by a user-defined calibration factor, see Eq. (A.22).

These transports are determined by generating velocity signals of the orbital velocities near the bed by using the Rienecker and Fenton (1981) method, see also Roelvink and Stive (1989). The (short wave) averaged sediment transport due to wave asymmetry, Equations (A.16) and (A.17), is determined by using the following averaging expressions of the near bed velocity signal (calibration coefficients included):

$$\langle u|u|^2 \rangle = FacA \langle \tilde{u}|\tilde{u}|^2 \rangle + 3 FacU \bar{u} \langle |\tilde{u}|^2 \rangle \tag{A.20}$$

$$\langle u|u|^3 \rangle = FacA \langle \tilde{u}|\tilde{u}|^3 \rangle + 4 FacU \bar{u} \langle |\tilde{u}|^3 \rangle \tag{A.21}$$

in which: $\tilde{u}$ is orbital velocity signal, $\bar{u}$ is averaged flow velocity (due to tide, undertow, wind, *etc.*), *FacA* is user-defined calibration coefficient for the wave asymmetry, *FacU* is user-defined calibration coefficient for the averaged flow.

The suspended transport relation due to the bed slope is implemented as:

$$S_{s;slope,\xi}(t) = \frac{\rho c_f \varepsilon_s}{(\rho_s - \rho) g (1 - p) w} \Big[\frac{\varepsilon_{sl}}{w} |u(t)|^5 \Big] \frac{\partial z_b}{\partial \xi} \tag{A.22}$$

where ε_{sl} is user-defined calibration coefficient *EpsSL*.

B. van Rijn transport formula (1993)

Van Rijn distinguishes between sediment transport below the reference height a, which is treated as bedload transport and that above the reference height which is treated as suspended-load. Sediment is entrained in the water column by imposing a reference concentration at the reference height.

The reference concentration is calculated in accordance with van Rijn (2000) as:

$$c_a^\ell = 0.015 \rho_s^{(\ell)} \frac{d_{50}^{(\ell)} (T_a^{(\ell)})^{1.5}}{a (D_*^{(\ell)})^{0.3}} \tag{A.23}$$

where $c_a^{(\ell)}$ mass concentration at the van Rijn's reference height a. In order to evaluate this expression the following quantities must be calculated: $D_*^{(\ell)}$ non-dimensional particle diameter, $T_a^{(\ell)}$ non-dimensional bed-shear stress, $\mu_c^{(\ell)}$ efficiency factor current, $f'^{(\ell)}_c$ gain related friction factor, $f_c^{(\ell)}$ total current-related friction factor, $\tau_{b,cw}$ bed shear stress due to current in the presence of waves, $\mu_w^{(\ell)}$ efficiency factor waves, $\tau_{b,w}$ bed shear stress due to waves, f_w total wave-related friction factor:

$$D_*^{(\ell)} = d_{50}^{(\ell)} \left[\frac{(s^{(\ell)} - 1)g}{v^2} \right]^{\frac{1}{3}} \quad (A.24)$$

$$T_a^{(\ell)} = \frac{(\mu_c^{(\ell)} \tau_{b,cw} + \mu_w^{(\ell)} \tau_{b,cw}) - \tau_{cr}^{(\ell)}}{\tau_{cr}^{(\ell)}} \quad (A.25)$$

$$\mu_c^{(\ell)} = \frac{f'^{(\ell)}_c}{f_c} \quad (A.26)$$

$$f'^{(l)}_c = 0.24 \left[log_{10} \left(\frac{12h}{3d_{90}^{(\ell)}} \right) \right]^{-2} \quad (A.27)$$

$$f_c^{(\ell)} = 0.24 \left[log_{10} \left(\frac{12h}{k_s} \right) \right]^{-2} \quad (A.28)$$

$$\tau_{b,cw} = \rho_w u_*^2 \quad (A.29)$$

$$\mu_w^{(\ell)} = \max\left(0.063, \frac{1}{8}\left(1.5 - \frac{H_s}{h}\right)^2\right) \quad (A.30)$$

$$\tau_{b,w} = \frac{1}{4}\rho_w f_w (\hat{U}_\delta)^2 \quad (A.31)$$

$$f_w = exp\left[-6 + 5.2\left(\frac{\hat{A}_\delta}{k_{s,w}}\right)^{-0.19}\right] \quad (A.32)$$

in which, h is water depth; $\hat{A}$ is peak orbital excursion at the bed, $\hat{A} = T_p \hat{U}_\delta / 2\pi$; $d_{50}^{(\ell)}$ is median sediment diameter, $d_{90}^{(\ell)}$ is 90% sediment passing size, $d_{90}^{(\ell)} = 1.5 d_{50}^{(\ell)}$; k_a is apparent bed roughness felt by the flow when waves are present, which is calculated by Delft3D-FLOW using the wave-current interaction formulation selected, $k_a \leq 10 k_s$; k_s is user-defined current-related effective roughness height (space varying); $k_{s,w}$ is wave related roughness, calculated from ripple height; u_z is velocity magnitude taken from a near-bed computational layer. In a current-only situation the velocity in the bottom computational layer is used. Otherwise, if waves are active, the velocity is taken from the layer closest to the height of the top of the wave mixing layer δ. δ_m is thickness of wave boundary mixing layer following van Rijn (1993), $\delta_m = 3\delta_w$ and $\delta_m \geqslant k_a$; δ_w is wave boundary layer thickness:

$$\delta_w = 0.0782 \hat{A}_\delta \left(\frac{\hat{A}_\delta}{k_{s,w}}\right)^{0.25} \quad (A.33)$$

$\hat{U}_\delta$ is peak orbital velocity at the bed; z_u is height above bed of the near-bed velocity (u_z) used in the calculation of bottom shear stress due to current; $\tau_{cr}^{(\ell)}$ is critical bed shear stress:

$$\tau_{cr}^{(\ell)} = (\rho_s^{(\ell)} - \rho_w) g d_{50}^{(\ell)} \theta_{cr}^{(\ell)} \quad (A.34)$$

in which, $\theta_{cr}^{(\ell)}$ is threshold parameter, calculated according to the classical Shields curve as modelled by van Rijn (1993) as a function of the non-dimensional grain size D_*.

Note in Eq. (A.29) that the bed shear velocity μ_* is calculated in such a way that van Rijn's wave-current interaction factor α_{cw} is not required. To avoid the need for excessive user input, the wave related roughness $k_{s,w}$ is related to the estimated ripple height, using the relationship:

$$k_{s,w} = RWAVE\Delta_r, \quad \Delta_r = 0.025 \quad (0.01 \text{ m} \leqslant k_{s,w} \leqslant 0.1 \text{ m}) \quad (A.35)$$

where, RWAVE is the user-defined wave roughness adjustment factor, recommended to be in the range 1 – 3, default 2. Δ_r is estimated ripple height.

The representative diameter of the suspended sediment $d_s^{(\ell)}$ generally given by the user-defined sediment diameter SEDDIA (d_{50} of bed material) multiplied by the user-defined factor FACDSS can be overruled in case the van Rijn (1993) transport formula is selected. This achieved by setting IOPSUS = 1 the representative diameter of the suspended sediment will then be set to:

$$d_s^{(\ell)} = \begin{cases} 0.64 d_{50}^{(\ell)} & \text{for } T_A^{(\ell)} \leqslant 1 \\ d_{50}^{(\ell)} (1 + 0.015(T_A^{(\ell)} - 25)) & \text{for } 1 < T_A^{(\ell)} \leqslant 25 \\ d_{50}^{(\ell)} & \text{for } 25 < T_A^{(\ell)} \end{cases} \quad (A.36)$$

where $T_a^{(\ell)}$ is non-dimensional bed-shear stress, given by above Eq. (A.25).

For simulations including waves the magnitude and direction of the bedload transport on a horizontal bed are calculated using an approximation method developed by van Rijn et al. (2003). The method computes the magnitude of the bedload transport as:

$$|S_b| = 0.006 \rho_s w_s d_{50}^{(\ell)} M^{0.5} M_e^{0.7} \quad (A.37)$$

in which, S_b is bed load transport (kg/m/s); M is sediment mobility number due to waves and currents; M_e is excess sediment mobility number, given by:

$$M = \frac{\nu_{off}^2}{(s-1)gd_{50}} \quad (A.38)$$

$$M_e = \frac{(\nu_{off} - \nu_{cr})^2}{(s-1)gd_{50}} \quad (A.39)$$

$$\nu_{off} = \sqrt{\nu_R^2 + U_{on}^2} \quad (A.40)$$

where, ν_{cr} is critical depth averaged velocity for initiation of motion (based on a parameterisation of the Shields curve, m/s); ν_R is magnitude of an equivalent depth-averaged velocity computed from the velocity in the bottom computational layer, assuming a logarithmic velocity profile (m/s); U_{on} is near-bed peak orbital velocity (m/s) in onshore direction (in the direction on wave propagation) based on the significant wave height. U_{on} (and U_{off} used below) are the high frequency near-bed orbital velocities due to short waves and are computed using a modification of the method of Isobe and Horikawa (1982). This method is a parameterisation of fifth-order Stokes wave theory and third-order cnoidal wave theory which can be used over a wide range of wave conditions and takes into account the non-linear effects that occur as waves propagate in shallow water (Grasmeijer and van Rijn, 1998).

The direction of the bedload transport vector is determined by assuming that it is composed of two parts: part due to current ($S_{b,c}$) which acts in the direction of the near-bed current, and part due to waves ($S_{b,w}$) which acts in the direction of wave propagation. These components are determined as follows

$$S_{b,c} = \frac{S_b}{\sqrt{1 + r^2 + 2|r|\cos\varphi}} \qquad (A.41)$$

$$|S_{b,w}| = r|S_{b,c}| \qquad (A.42)$$

$$r = \frac{(|U_{on}| - v_{cr})^3}{(|v_R| - v_{cr})^3} \qquad (A.43)$$

where, $S_{b,w} = 0(r < 0.01)$, $S_{b,c} = 0(r > 100)$, and φ is angle between current and wave direction for which van Rijn (2003) suggests a constant value of 90°.

Also included in the "bedload" transport vector is an estimation of the suspended sediment transport due to wave asymmetry effects. This is intended to model the effect of asymmetric wave orbital velocities on the transport of suspended material within about 0.5m of the bed (the bulk of the suspended transport affected by high frequency wave oscillations). This wave-related suspended sediment transport is again modelled using an approximation method proposed by van Rijn (2001):

$$S_{s,w} = f_{\text{SUSW}} \gamma U_A L_T \qquad (A.44)$$

$$U_A = \frac{U_{on}^4 - U_{off}^4}{U_{on}^3 + U_{off}^3} \qquad (A.45)$$

$$L_T = 0.007 \rho_s d_{50} M_e \qquad (A.46)$$

in which, $S_{s,w}$ is wave-related suspended transport (kg/m/s); f_{SUSW} is user-defined tuning parameter; γ is phase lag coefficient (0.2); U_A is velocity asymmetry value (m/s); L_T is suspended sediment load (kg/m^2).

The three separate transport modes are imposed separately. The direction of the bedload due to currents $S_{b,c}$ is assumed to be equal to the direction of the current, whereas the two wave related transport components $S_{b,w}$ and $S_{s,w}$ take on the wave propagation direction. This results in the following transport components

$$S_{bc,u} = \frac{u_{b,u}}{|u_b|}|S_{b,c}|, \; S_{bc,v} = \frac{u_{b,v}}{|u_b|}|S_{b,c}| \qquad (A.47)$$

$$S_{bw,u} = S_{b,w}\cos\varphi, \; S_{bw,u} = S_{b,w}\cos\varphi \qquad (A.48)$$

$$S_{sw,u} = S_{s,w}\cos\varphi, \; S_{sw,u} = S_{s,w}\sin\varphi \qquad (A.49)$$

where φ is the local angle between the direction of wave propagation and the computational grid. The different transport components can be calibrated independently by using the keywords *Bed*, *BedW* and *SusW* in the morphology input file.

C. Figures

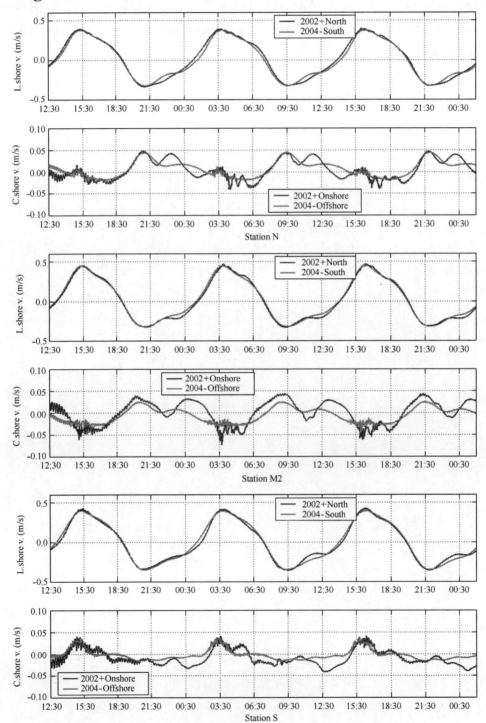

Fig. A. 1 Comparison of Water levels, longshore and cross-shore velocities at Station N, M2 and S. No waves present. The stations are shown in Fig. 4. 3.

Appendix

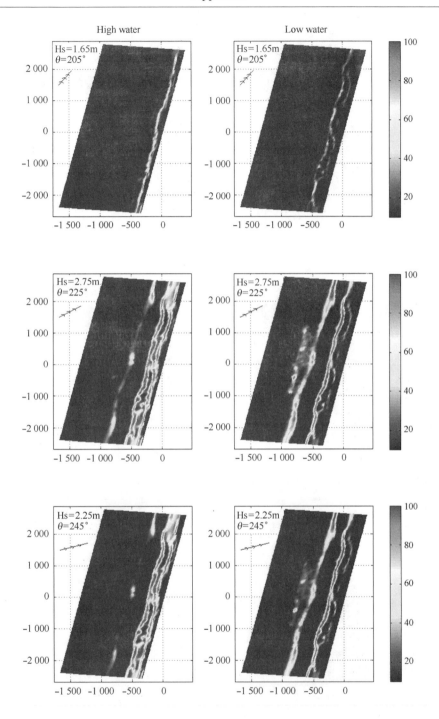

Fig. A.2 Energy dissipation rate of the waves coming from the southwest at high water and low water (Unit: N/m/s)

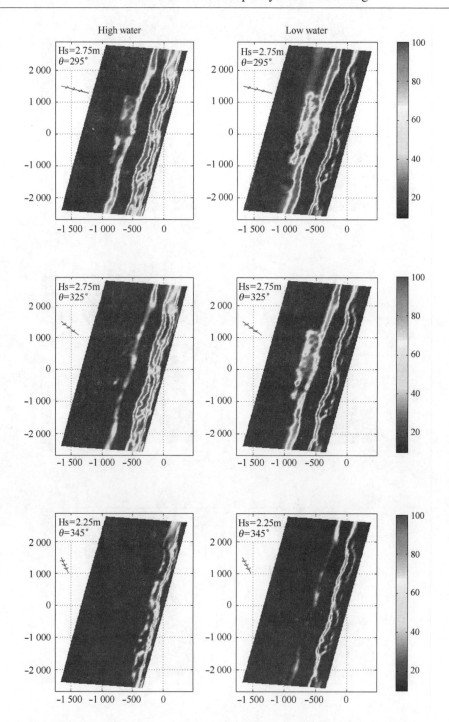

Fig. A. 3 Energy dissipation rate of the waves coming from the northwest at high water and low water (Unit: N/m/s)

Appendix

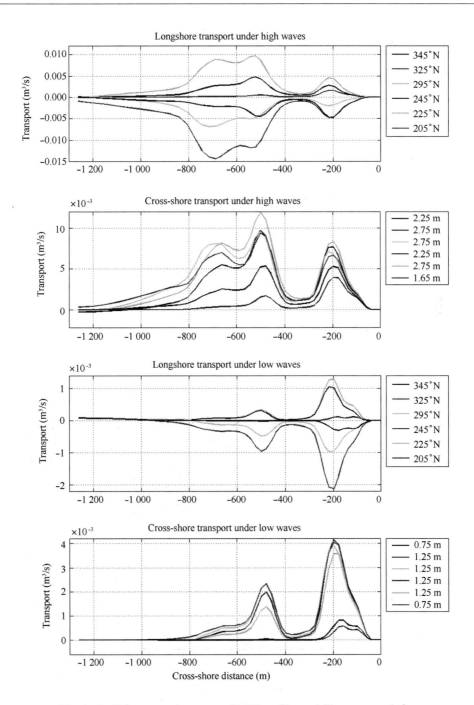

Fig. A.4 Tide-averaged tranport of 1DH profile modelling case psw1p0. Positive means northward longshore transport or landward cross-shore transport. Negative means southward longshore transport or seaward cross-shore transport.

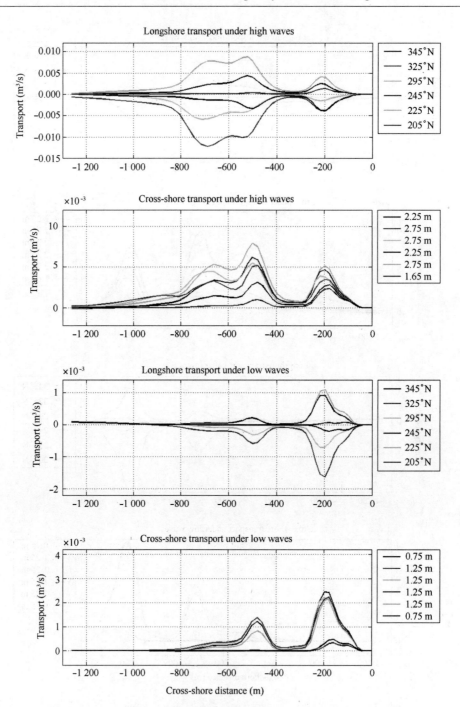

Fig. A. 5 Tide-averaged transport of 1DH profile modelling case psw0p5. Positive means northward longshore transport or landward cross-shore transport. Negative means southward longshore transport or seaward cross-shore transport.

Appendix

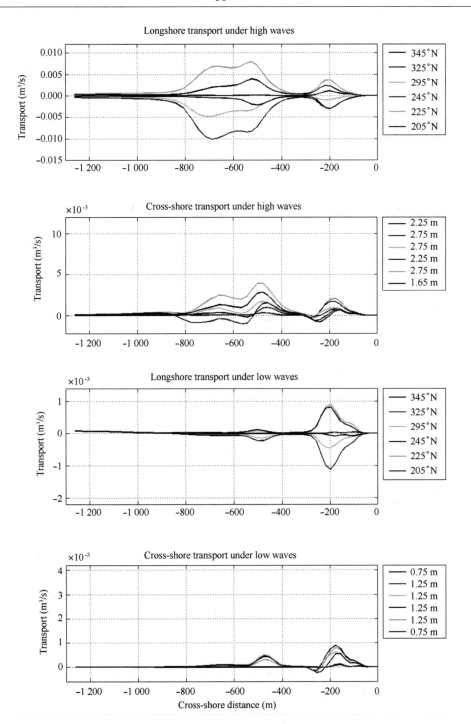

Fig. A. 6 Tide-averaged transport of 1DH profile modelling case psw0p0.
Positive means northward longshore transport or landward cross-shore transport.
Negative means southward longshore transport or seaward cross-shore transport.

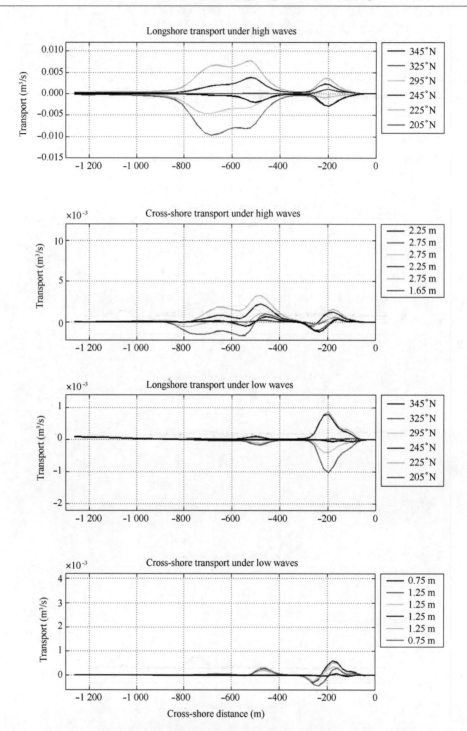

Fig. A. 7 Tide-averaged transport of 1DH profile modelling case pbw0p5.
Positive means northward longshore transport or landward cross-shore transport.
Negative means southward longshore transport or seaward cross-shore transport.

Appendix

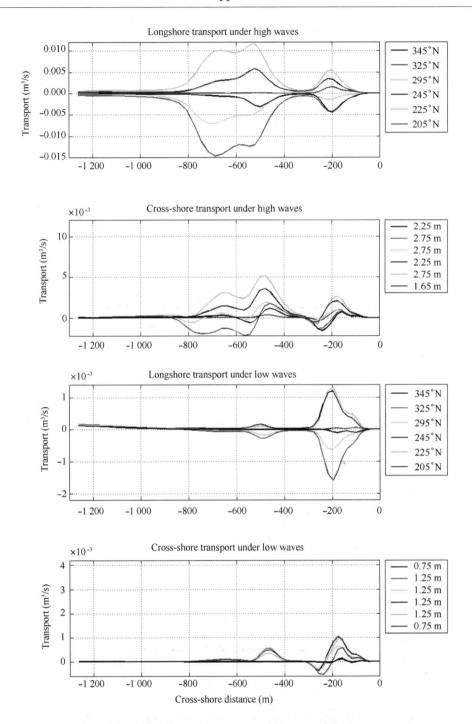

Fig. A. 8 Tide-averaged transport of 1DH profile modelling case ps1p5.
Positive means northward longshore transport or landward cross-shore transport.
Negative means southward longshore transport or seaward cross-shore transport.

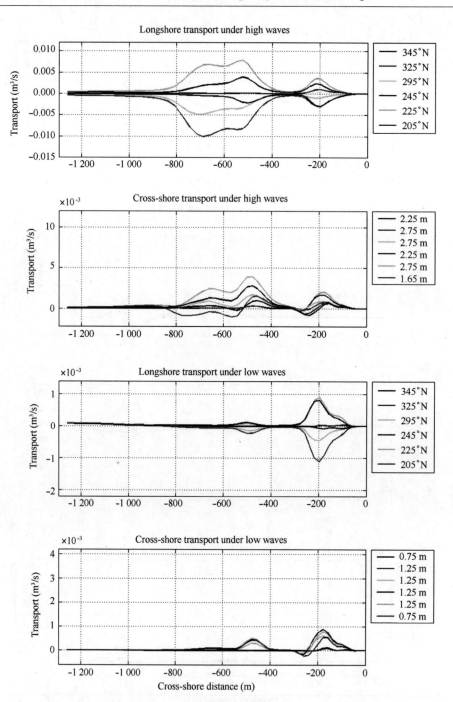

Fig. A. 9 Tide-averaged transport of 1DH profile modelling case pb0p5. Positive means northward longshore transport or landward cross-shore transport. Negative means southward longshore transport or seaward cross-shore transport.

Appendix

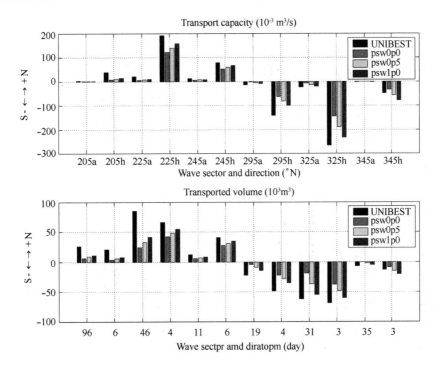

Fig. A. 10 Tide-averaged and cross-section integrated longshore transport capacity and transported volume accumulated over morphological duration per wave group of 1DH profile model for different f_{SUSW}. The results of UNIBEST is directly taken from the previous study (van Duin, 2002).

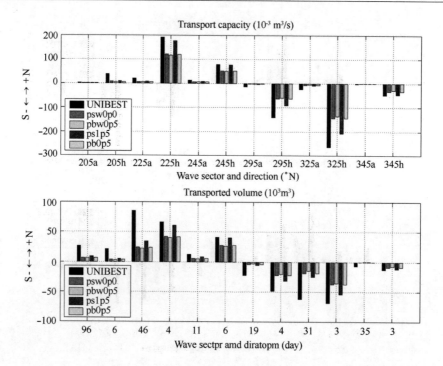

Fig. A.11　Tide-averaged and cross-section integrated longshore transport capacity and transported volume accumulated over morphological duration per wave group of 1DH profile model for different calibrating factors. The results of UNIBEST is directly taken from the previous study (van Duin, 2002).

Appendix

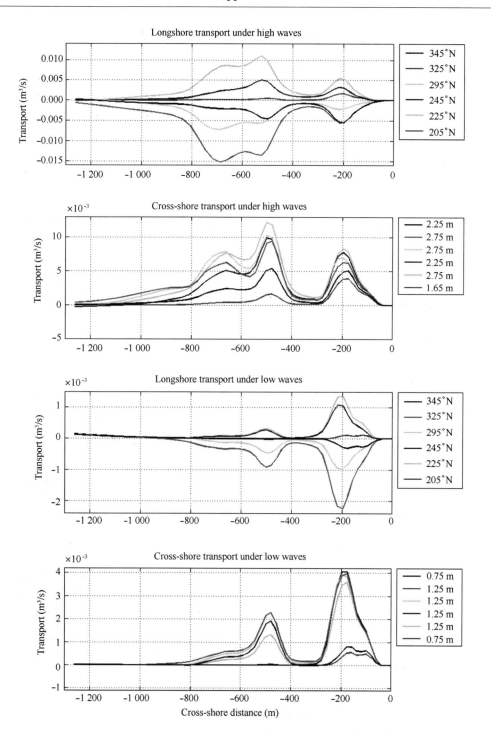

Fig. A. 12 Tide-averaged transport of 2DV profile modelling case psw1p0.
Positive means northward longshore transport or landward cross-shore transport.
Negative means southward longshore transport or seaward cross-shore transport.

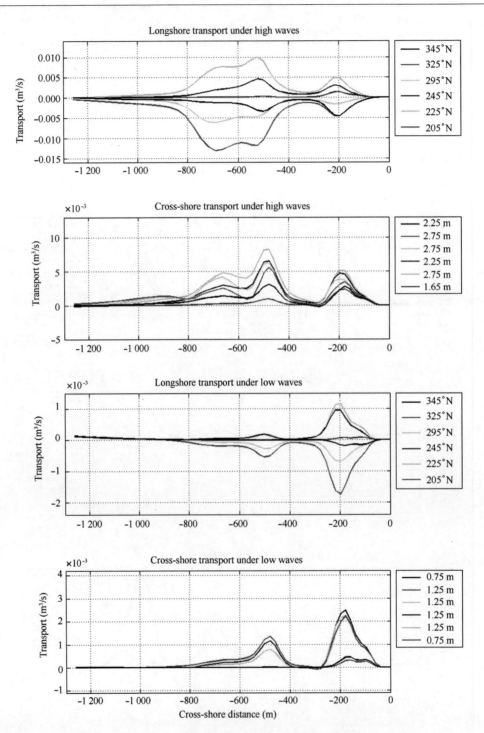

Fig. A. 13 Tide-averaged transport of 2DV profile modelling case psw0p5.
Positive means northward longshore transport or landward cross-shore transport.
Negative means southward longshore transport or seaward cross-shore transport.

Appendix

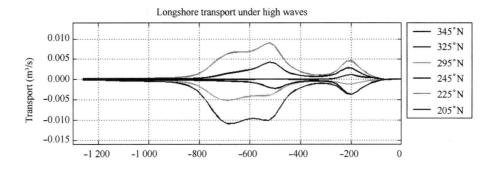

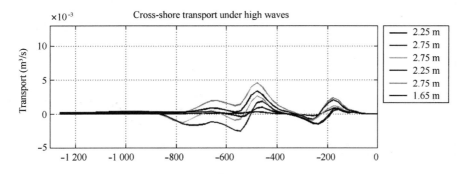

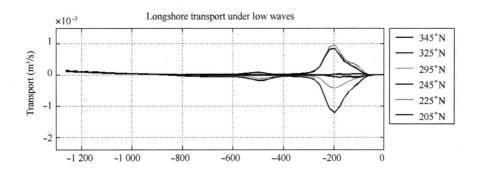

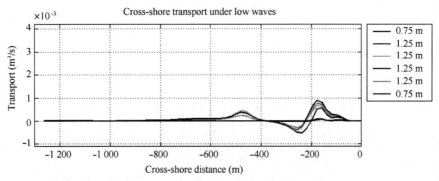

Fig. A. 14   Tide-averaged transport of 2DV profile modelling case psw0p0.

Positive means northward longshore transport or landward cross-shore transport.

Negative means southward longshore transport or seaward cross-shore transport.

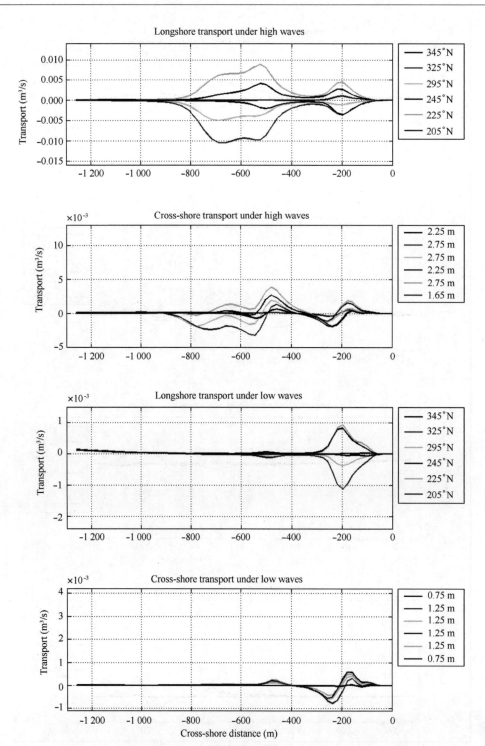

Fig. A. 15 Tide-averaged transport of 2DV profile modelling case pbw0p5. Positive means northward longshore transport or landward cross-shore transport. Negative means southward longshore transport or seaward cross-shore transport.

Appendix

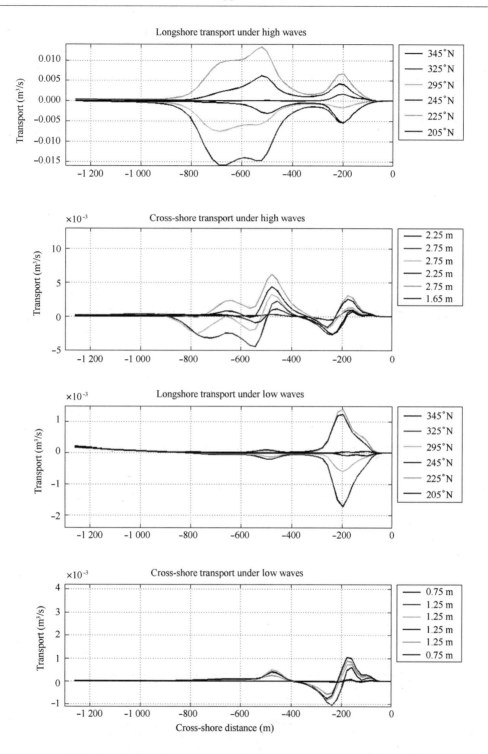

Fig. A. 16 Tide-averaged transport of 2DV profile modelling case ps1p5. Positive means northward longshore transport or landward cross-shore transport. Negative means southward longshore transport or seaward cross-shore transport.

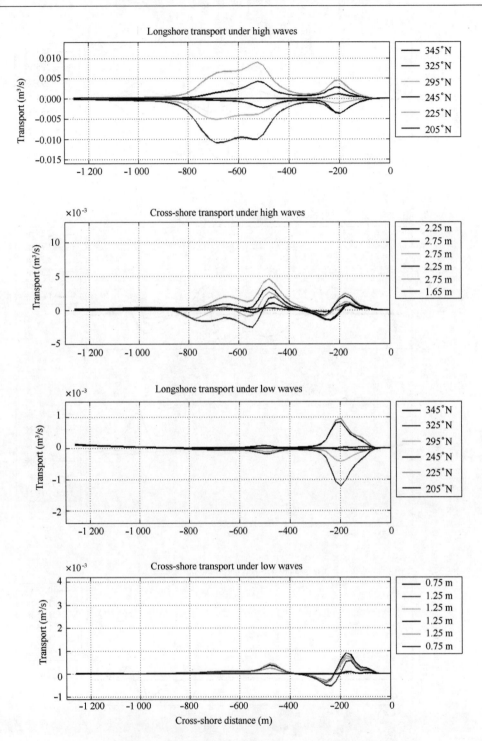

Fig. A. 17 Tide-averaged transport of 2DV profile modelling case pb0p5. Positive means northward longshore transport or landward cross-shore transport. Negative means southward longshore transport or seaward cross-shore transport.

Appendix

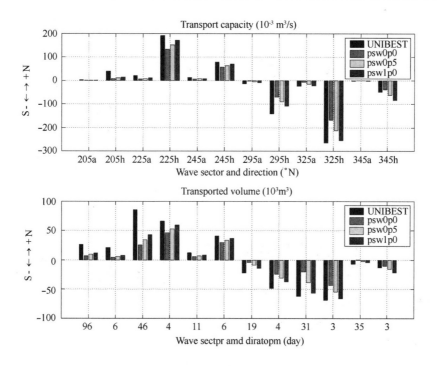

Fig. A. 18 Tide-averaged and cross-section integrated longshore transport capacity and transported volume accumulated over morphological duration per wave group of 2DV profile model for different f_{SUSW}. The results of UNIBEST is directly taken from the previous study (van Duin, 2002).

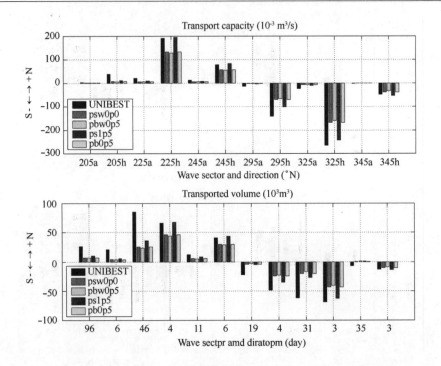

Fig. A. 19 Tide-averaged and cross-section integrated longshore transport capacity and transported volume accumulated over morphological duration per wave group of 2DV profile model for different calibrating factors. The results of UNIBEST is directly taken from the previous study (van Duin, 2002).